SUPER ARDUINO

For general information on our other products and services or to obtain technical support, please contact our Customer Care Department within the United States at (866) 744-2665, or outside the United States at (510) 253-0500.

Rockridge Press publishes its books in a variety of electronic and print formats. Some content that appears in print may not be available in electronic books, and vice versa.

Interior and Cover Designer: Michael Cook
Photo Art Director: Michael Hardgrove
Editor: David Lytle
Production Editor: Kurt Shulenberger
Photography: © 2019 Kenneth Hawthorn
ISBN: Print 978-1-64152-599-2 | eBook 978-1-64152-600-5
Ro

SUPER ARDUINO

Step-by-Step Instructions to Build Cool Gadgets

KENNETH HAWTHORN

FOR APRIL

CONTENTS

ARDUINO: A MICROCONTROLLER FOR EVERYONE

You are not currently flying around in your flying suit because, well, you have not yet built a flying suit. Over the last few years artists, tinkerers, students, and inventors have seen more of their visions come to life because of the inexpensive microcontroller known as Arduino. A microcontroller is a very simple computer that runs a program that can control things such as lights or motors. With this tiny device, an artist can create a light installation with millions of colors that react to the people looking at the installation. It allows a robotic kitten to judge the distance to an obstacle. It will provide opportunities to those who feel compelled to attach small jet engines to a wearable wing to actually fly. This book is a journey through several Arduino projects that will enable you to build a skill set to tackle your own dream projects. I wrote this book so you can build a machine that will make you smile.

The projects in this book are designed to be completed in order. There is nothing stopping you from building any project within this book as a stand-alone project, but the order of projects was carefully thought out to build the reader's confidence across the broad spectrum of devices that can be used with Arduino.

The way to understand Arduino is to look at the circuit board as an "if this happens, then do that" machine. If an input happens, then an output occurs. Typically, Arduino is looking at some kind of input event (such as a switch being pressed or a countdown reaching zero) and using the input event as a programmed trigger for an output event (such as a motor turning on or an LED illuminating).

Arduino and You

MATERIALS USED IN CHAPTER 1:

- **Arduino Uno R3Board**
- **USB Type 2.0 Printer Cable**
- **Laptop Computer**

Listed above are the materials that will be used in this chapter. The first thing that you will notice when shopping online for an Arduino is the variety of circuit boards that appear in the search. This is because the folks who invented Arduino (www.arduino.cc) made the incredibly generous decision to freely release their schematics and design data as open-source hardware and software. As a result, everyone is making his or her own version of the microcontroller and incorporating the name "Arduino" due to the origin of the design. This results in a plethora of options and styles being available.

Figure 1.1
Pictured above is an ultrasonic distance sensor (left) that can act as an input device for the Arduino and a servo motor (right) that can act as an output device.

Input	→	Arduino	→	Output
Button Press	→	Arduino	→	Motor On
Timer On	→	Arduino	→	LED On

START WITH THE OLD-SCHOOL CHIP

To add to the confusion in choosing your first board, there are several variants of Arduino designs. Technically, we are looking for the variant named an Arduino Uno. Other variants are called the Arduino Leonardo, YUN, and Mega, to name just a few of the 20 most commonly found. The Uno was the first design released and is by far the most commonly used model. Below are two copies of the Uno that I have used with success in my classroom over the past five years. The most reliable boards, other than an authentic Arduino ($20), come from Elegoo and SunFounder and cost approximately $5 to $7. If you are an Amazon user, I would suggest typing in "Arduino Uno R3" and choosing any starter kit that is rated four stars or higher. The only caveat I can add is to look at the picture and make sure that the microchip on the board looks similar to the one in figure 1.2. This is because some manufacturers are using microchips that emulate the functionality of the original chip, which sometimes require the installation of a separate software driver on your computer.

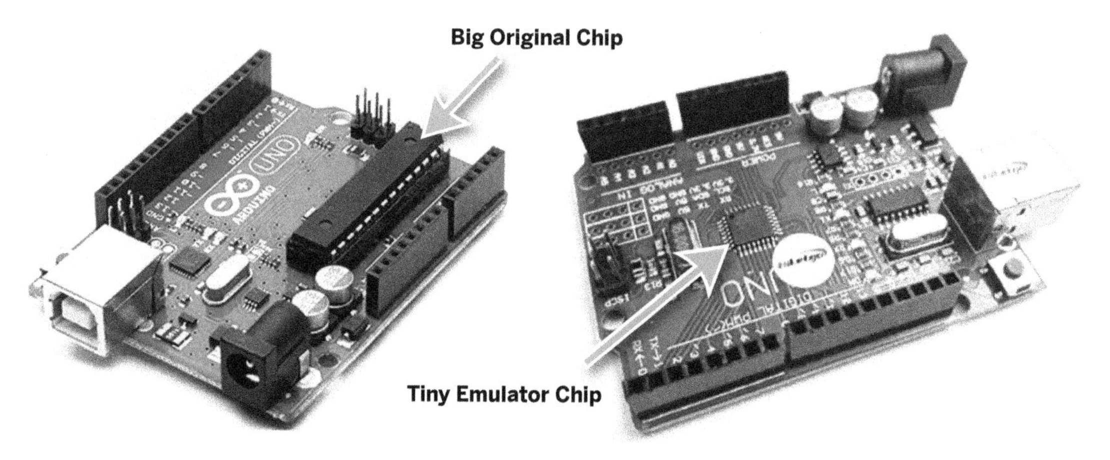

Figure 1.2

Pictured above are two different types of Arduino Uno boards. Both should function identically, but the version containing the tiny emulator may require extra steps to install a software driver on some laptops in order for the board to work correctly. When in doubt, look for the big, bulky microchip.

GETTING STARTED

We will begin our adventure by becoming familiar with the basics of using an Arduino. A company called Autodesk has developed a wonderful tool that will allow us to explore the basic functionality of an Arduino with an online simulator. This is a great way to establish a basic understanding regarding the steps necessary to write code for your Arduino and to simulate a successful circuit. Next, we will explore setting up your laptop to assist with building that same circuit while using the physical Arduino board. To get started, navigate to www.Tinkercad.com.

1. Go to www.Tinkercad.com and click "Join Now."
2. Create an account.

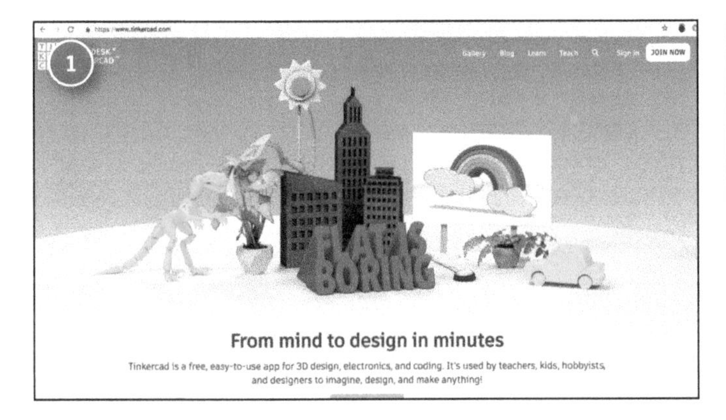

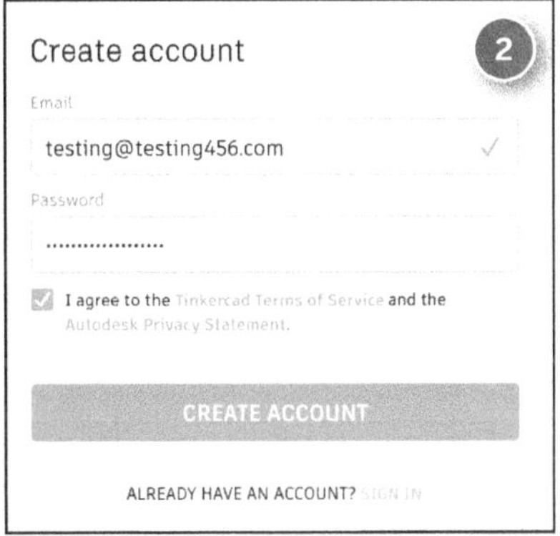

3. Click "Circuits."

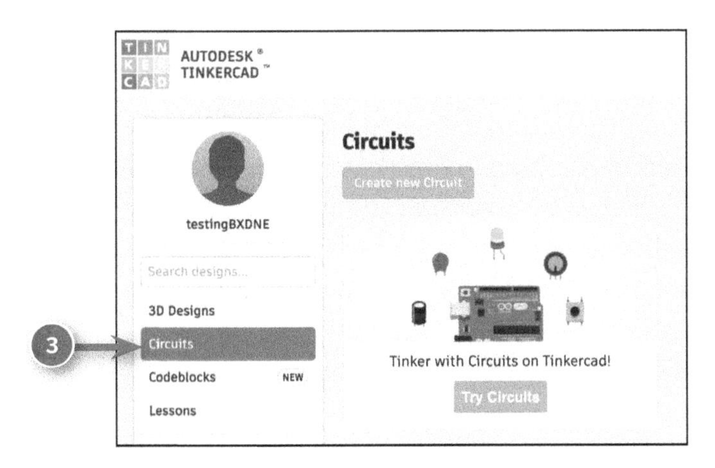

4. Type "Arduino" under components and select "Blink" under Starters.
5. Play around with these buttons.

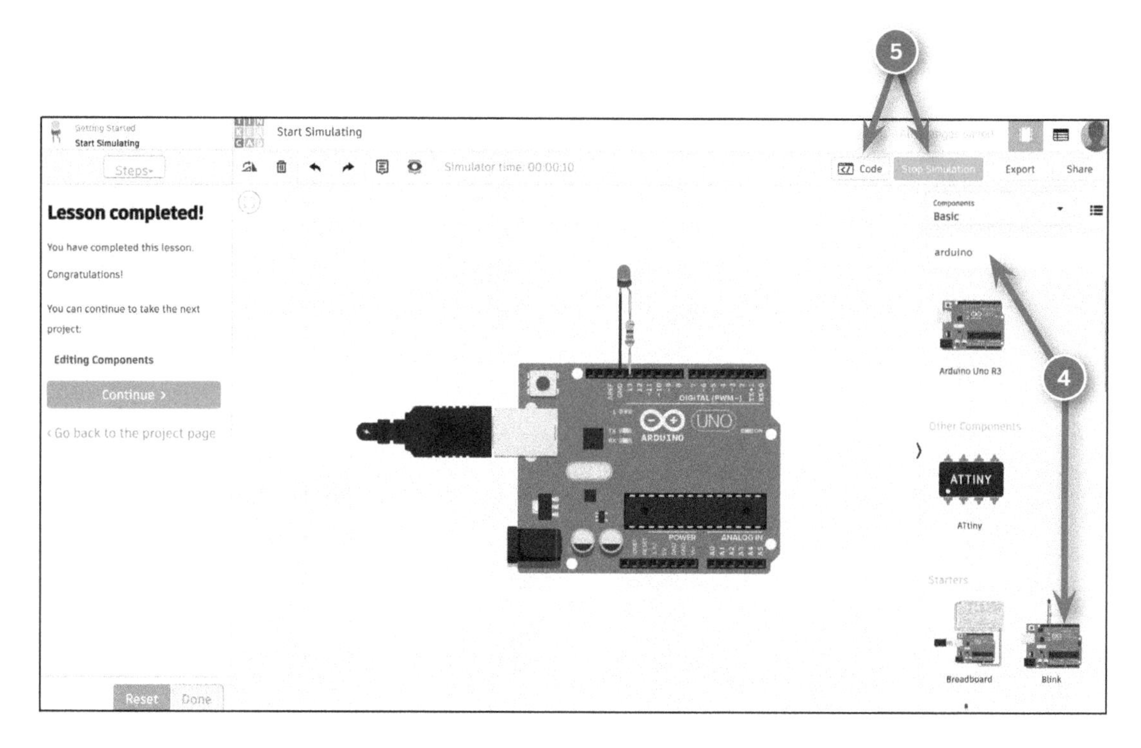

Once you have the "Blink" example program up and running, we are ready to start with the physical Arduino board.

YOUR BOARD: KNOW IT INSIDE AND OUT

I strongly dislike overly technical explanations when simple words will suffice. Therefore, my goal in this section is for you to gain a basic understanding of how and why the Arduino microcontroller is laid out the way that it is. Earlier, I described Arduino as the part of the machine that lives between inputs and outputs. Figure 1.3 on page 5 builds on that understanding.

Figure 1.3

1. A place to connect a battery pack for when your Arduino is not plugged into your computer
2. USB cable goes here with the other end in your computer when being programmed
3. Reset Button
4. Power for the project inputs and outputs you connect to Arduino
5. Signals from inputs go here
6. Microchip that stores code and directs decisions about inputs and outputs
7. Signals to control outputs come out here
8. Built-in LED for testing

As complicated as some people might like to make electronics sound, the truth is a microcontroller is simply a board that gives us the ability to enter inputs and outputs onto a chip that stores and runs the programs that we write.

INSTALLING AND USING THE ARDUINO INTERFACE

We will need an interface to program our Arduino. Programmers call this interface an Integrated Development Environment (IDE). There are four options to choose from. The first three options depend on the kind of laptop you use. The fourth option available to you is to use an online interface that is platform agnostic, which means that it does not matter what brand of laptop you work on. Take a moment to review the pros and cons of each option below, and choose the one that makes sense to you.

#1 Arduino on Windows

What Works: The Arduino IDE on a Windows machine is stable and backward compatible with every version of Windows since XP. You can also choose to download a Windows app instead of a program for those of you who like a Windows-10-style tablet experience.

Frustrations: This is Windows, land of the forced update and missing software drivers. Expect to spend time chasing down bugs when the USB port does not show up correctly.

#2 Arduino on Mac

What Works: The Arduino IDE on a Mac is simple. You can even choose to run the program directly from the downloads folder in the Finder window. The betas (early preview editions of new versions of the IDE) are usually stable, and I have never gone wrong downloading the most current version of the Arduino IDE.

Frustrations: Users should be reasonably good at using the Finder window. Each Arduino program you run will need its own individual folder. It's best to avoid moving files out of their assigned folders. As of this writing the most recent Mac OS "Catalina" is breaking all 32 bit programs and is something to stay away from.

#3 Arduino on Chromebook

What Works: The Arduino IDE on a Chromebook is a very lightweight install that does not take up much memory or require much processing power, which can be an issue with other programs on Chromebook. This is a great option for individuals who are already comfortable on Chromebook.

Frustrations: As with many Chrome apps, you need a reliable Internet connection.

#4 Arduino Online

What Works: The Arduino Web editor does a great job simulating the download versions as a browser plug-in. This is the way to get Arduino to work on any computer and to be able to save and access your work from more than one computer.

Frustrations: Unlike a downloaded IDE, you are totally dependent on an Internet connection at all times when programming your Arduino. You are also trusting the programs you write to an online service that can go down or lose your files.

Installing Arduino on Windows

1. Start by going to www.arduino.cc.
2. Select "Software" (#1) and then "Downloads" (#2).

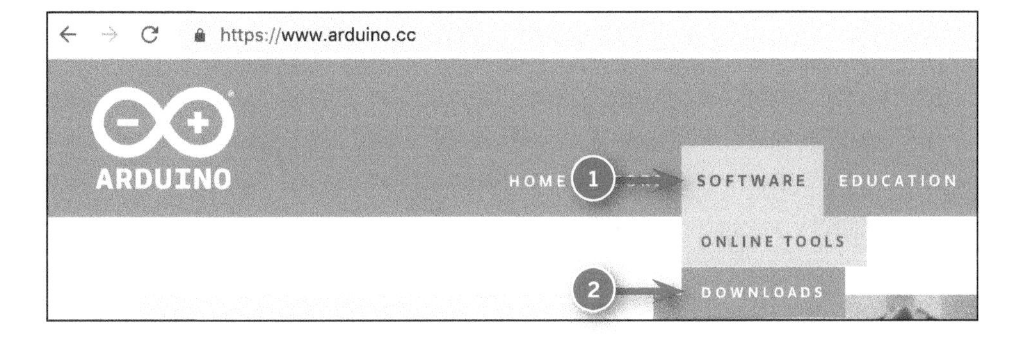

3. Click either the link for the traditional program on Windows machines running an operating system XP or newer (#3), or choose to download an app for Windows 8.1 or Windows 10 (#4). Do not be concerned with the version number of the Arduino IDE. That changes quite often, and newer versions are always backward compatible with older versions.

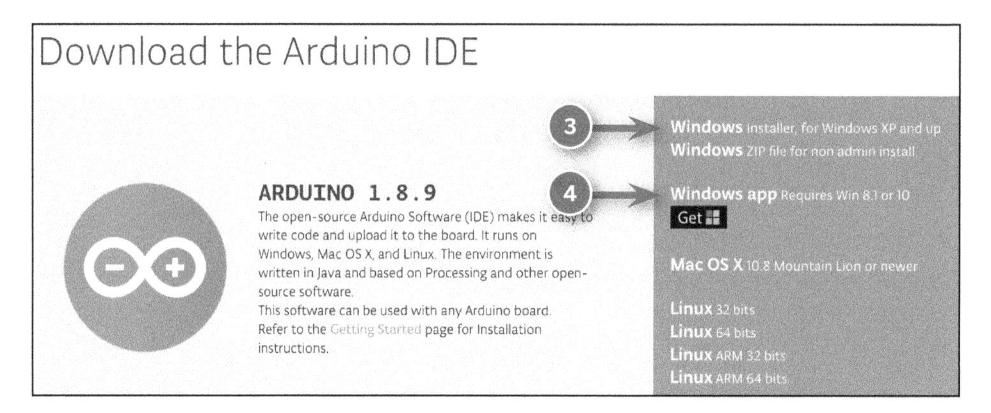

4. Allow access for the program to run in the Windows Defender pop-ups (#5).

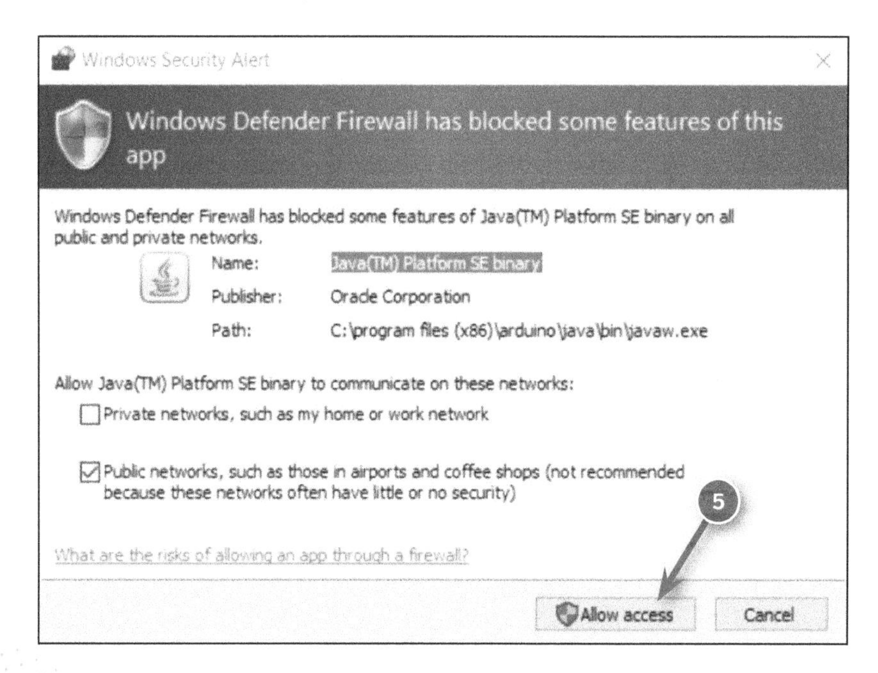

Installing Arduino on a Mac

1. Start by going to www.arduino.cc.
2. Select "Software" (#1) and then "Downloads" (#2).

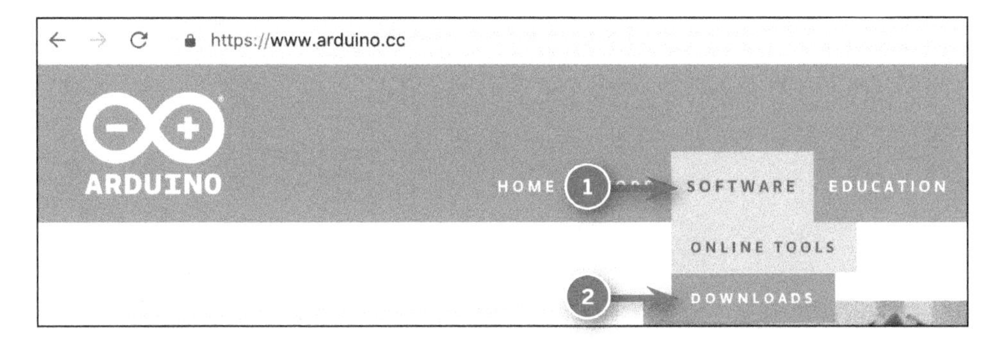

3. Click the link for OS X Mountain Lion or newer (#3). Do not be concerned with the version number of the Arduino IDE. That changes quite often, and newer versions are always backward compatible with older versions.
4. Open the downloaded file; it should unzip in less than 30 seconds (#4).

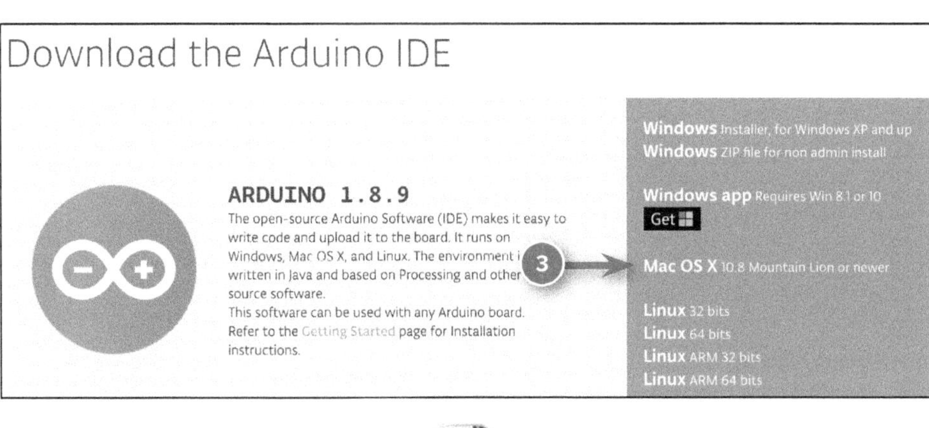

arduino-1.8.9-macosx (1).zip

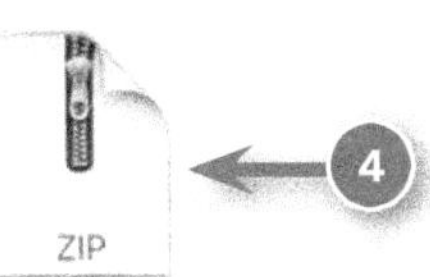

5. Locate the unzipped file in "Downloads" and open the program (**#5** and **#6**).

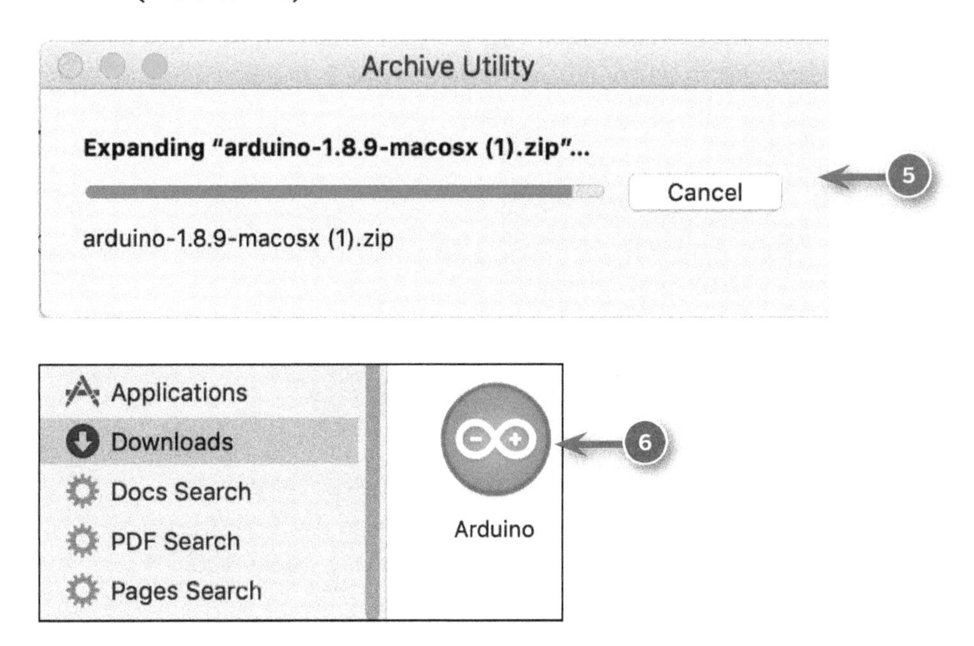

Installing Arduino on a Chromebook

1. Open the Chrome Web Store app on your Chromebook.
2. Search for "Arduino Create."
3. Click on the "Launch App" button" (#1).

4. At the icon for Arduino Create, tap the "App Launch" button (#2).

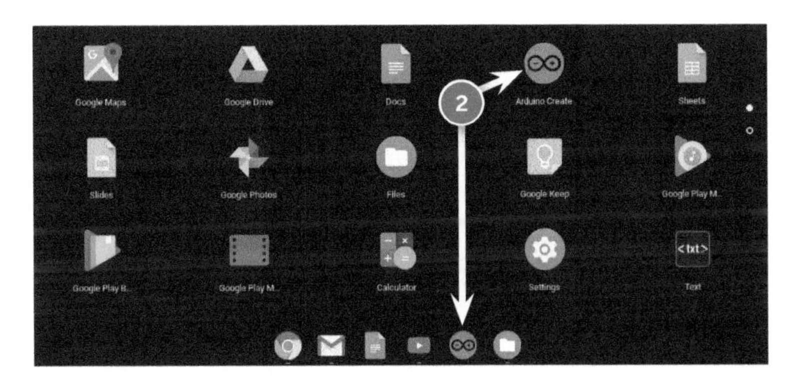

5. Double-click the Arduino icon, and it will open a Chrome browser tab (#3).

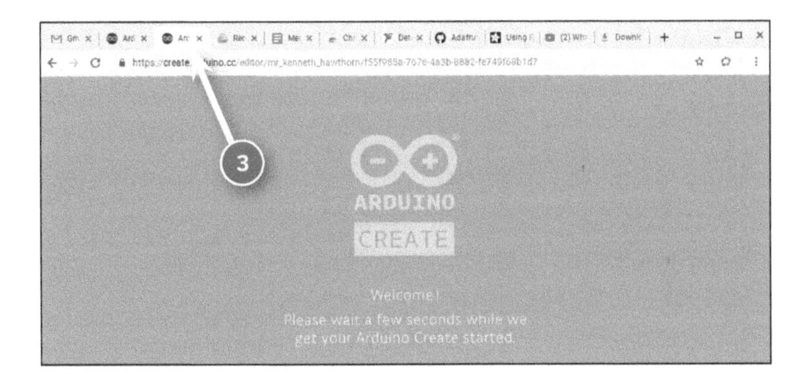

Arduino with the Online Web Editor

1. Open a web browser.
2. Navigate to www.Arduino.cc.
3. Click on "Software."
4. Click on "Online Tools."
5. Click on "Arduino Web Editor."
6. Click "Create New Account."
7. Open the Arduino Web Editor and click "New Sketch."

UNDERSTAND AND UPLOAD YOUR FIRST CODE

Regardless of the Arduino IDE you have chosen to use for your first project, all of the interfaces have exactly the same function and contain very similar layouts. For this book, I have chosen to use the Arduino IDE installed on a Mac. Everything that we cover will be almost identical in the Chromebook, Windows, or online editor options.

When you are working within Arduino, the formal name that is used for a program that you write is called "Sketch." In the screenshot below you can see how a blank sketch opens within the Arduino IDE. This is similar to opening up a new Microsoft Word document. I have labeled the most important parts of the Arduino IDE.

At this point I expect that the Arduino IDE will look completely foreign to you and may even be a little bit intimidating. However, I assure you that you will soon be capable of understanding the language. Your level of intimidation will dramatically decrease as we move along.

Inside this window you will see:

- A Check button that you can press to check your code for errors (#1).
- An arrow button, which is an icon that will upload your code from the Arduino IDE to the physical Arduino Uno board (#2).
- The Arduino sketch name (#3). Every Arduino sketch automatically names itself "sketch_" and the date it was created.

- The first section of code (**#4**), which addresses things that your program will do just once at the startup.
- The second section of code (**#5**), which contains commands that you would like to keep doing as long as the program is running.

  Note that each of these two sections of code are enclosed in a pair of curly brackets {}.

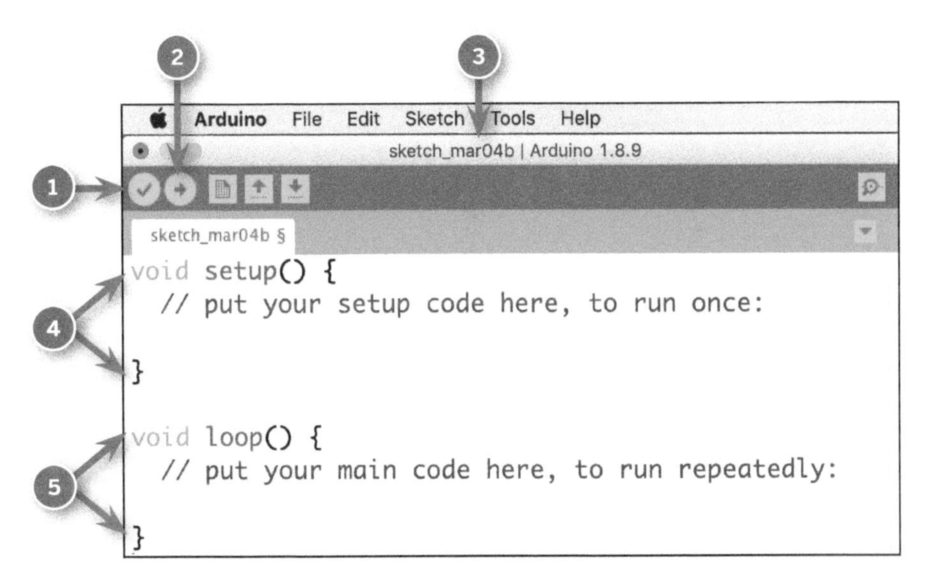

Note that the word "void" appears in a teal color, while the words "setup" and "loop" appear in a dark green. This is significant because the Arduino IDE changes the colors of the word you type as it interprets the meaning of each word. In the same way that the English language has verbs, nouns, and adjectives, in programming languages there are words that are treated with different purposes. Functions and variables, along with a host of other special words, are treated in a similar manner. For now, just know that the word "void" is usually associated with a function, so it appears in teal while the words "setup" and "loop" denote that the function will occur once or more than once.

Do not worry too much about this. For now it is enough to know that the Arduino IDE will change the color of the text in order to clarify how the words that you are typing will be interpreted in the sketch you eventually upload to your Arduino microcontroller.

UPLOADING A SKETCH

Let's get comfortable with the microcontroller by utilizing one of the sample sketches (programs) that the Arduino IDE provides. To do this, click on:
File > Tools > Examples > Basics > Blink.

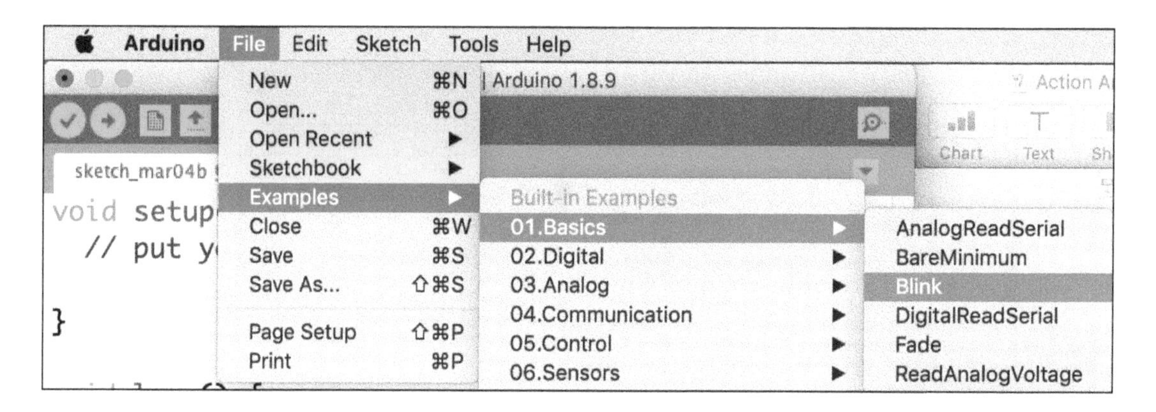

Arduino IDE example sketches are actually very good starting points for your own projects. There is no shame in modifying something that works well to use it for your own purposes. The example sketches also do a great job of covering advanced concepts. Nine years after picking up my first Arduino board, I still find myself visiting some of these examples to consider how they may be paired or used with my current projects.

Blink | Arduino 1.8.9

Blink

```
// the setup function runs once when you press reset or power the board
void setup() {
  // initialize digital pin LED_BUILTIN as an output.
  pinMode(LED_BUILTIN, OUTPUT);
}

// the loop function runs over and over again forever
void loop() {
  digitalWrite(LED_BUILTIN, HIGH);   // turn the LED on (HIGH is the voltage level)
  delay(1000);                       // wait for a second
  digitalWrite(LED_BUILTIN, LOW);    // turn the LED off by making the voltage LOW
  delay(1000);                       // wait for a second
}
```

The example sketch below has two more colors for words. The color orange denotes action words (similar to verbs) to change a value or delay an amount of time (in this case 1000 milliseconds, or one second). The gray text always appears to the right side of a double forward slash (//). This gray text is ignored by the computer and allows us to leave ourselves notes next to each line of code. These comments are valuable when we share code with friends or need to make notes pertaining to lines of code within the sketch.

Each Arduino has a series of holes called "pins." Pins are where you place wires to connect input and output signals. A closer look at figure 1.3 on page 5 will reveal holes running in two vertical rows on the left and right side of the Arduino. Note that these holes are all labeled with numbers or letters. Each of those is a "pin." The code above is flipping one of those pins that already has an LED attached to it to function only as output and blink on (high) and off (low) once every 1000 milliseconds (one second).

The next step is to check the code for errors and then upload it to the Arduino board. All we have to do to check our code is to click the check mark button (#1). However, before we hit the upload button (#2), we need to connect the Arduino board to the computer.

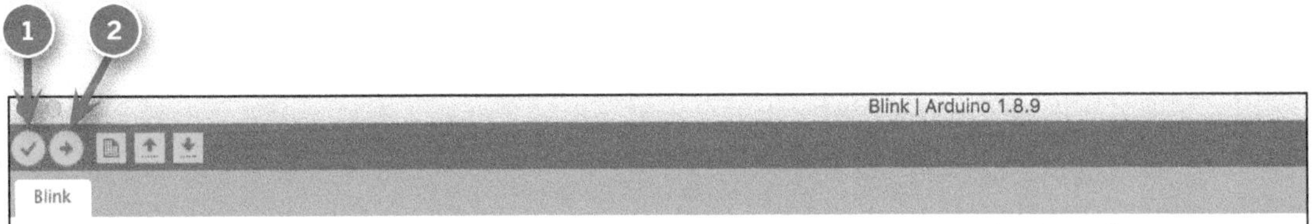

1 2

Blink | Arduino 1.8.9

Blink

```
// the setup function runs once when you press reset or power the board
void setup() {
  // initialize digital pin LED_BUILTIN as an output.
  pinMode(LED_BUILTIN, OUTPUT);
}

// the loop function runs over and over again forever
void loop() {
  digitalWrite(LED_BUILTIN, HIGH);   // turn the LED on (HIGH is the voltage level)
  delay(1000);                       // wait for a second
  digitalWrite(LED_BUILTIN, LOW);    // turn the LED off by making the voltage LOW
  delay(1000);                       // wait for a second
}
```

At this point we want to connect the flat and wide side of the USB cable (#1) to the computer and the more rectangular side of the USB cable (#2) to the Arduino Uno board, as shown below.

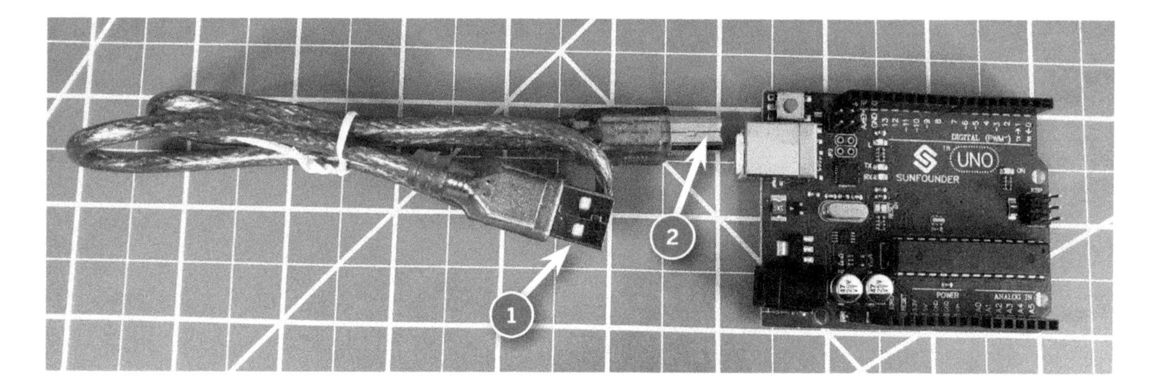

Once you have selected your Arduino Uno board, select Tools > Port and select USB and then click the 2) upload button. Once you have uploaded your code, you will see an LED blinking on the Uno (see page 18).

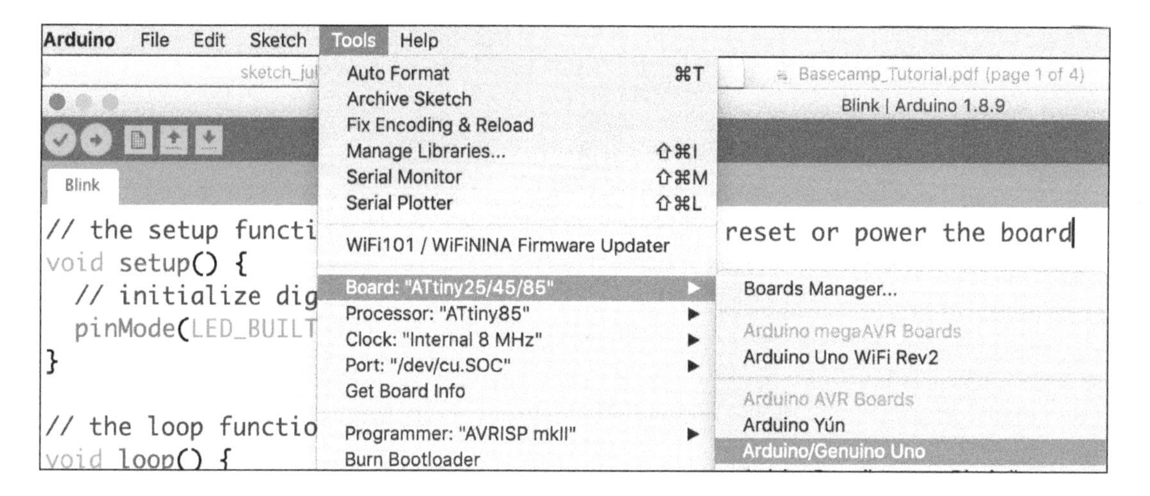

IN THIS CHAPTER YOU:

- learned what an Arduino Uno is.
- discovered what an IDE is and how to set up any computer to run the Arduino IDE.
- learned how to upload code from the Arduino IDE into the Arduino microcontroller.
- learned why open-source hardware brings down the cost of projects.
- how to use the Arduino simulator in Tinkercad to test circuits and code before you build.

TAKING IT FURTHER

In chapter 3 we will spend more time hooking up inputs and outputs to the pins on the Arduino Uno. We can take a quick preview of that right now by simply adding an external LED as an output device without writing any additional code. Simply select any 2.5- to 5-volt LED and identify the long leg and short legs. The long leg of the LED is where electricity comes in from an Arduino pin (positive), and the short leg is where the electricity flows out again to the negative side of the circuit. (I would likely be in trouble with any electrical engineer for that explanation, but I am writing this book for you, not for electrical engineers.) In order to get the external LED to blink beside the built-in LED, place the long leg into pin #13 and the short leg into GND just above pin #13.

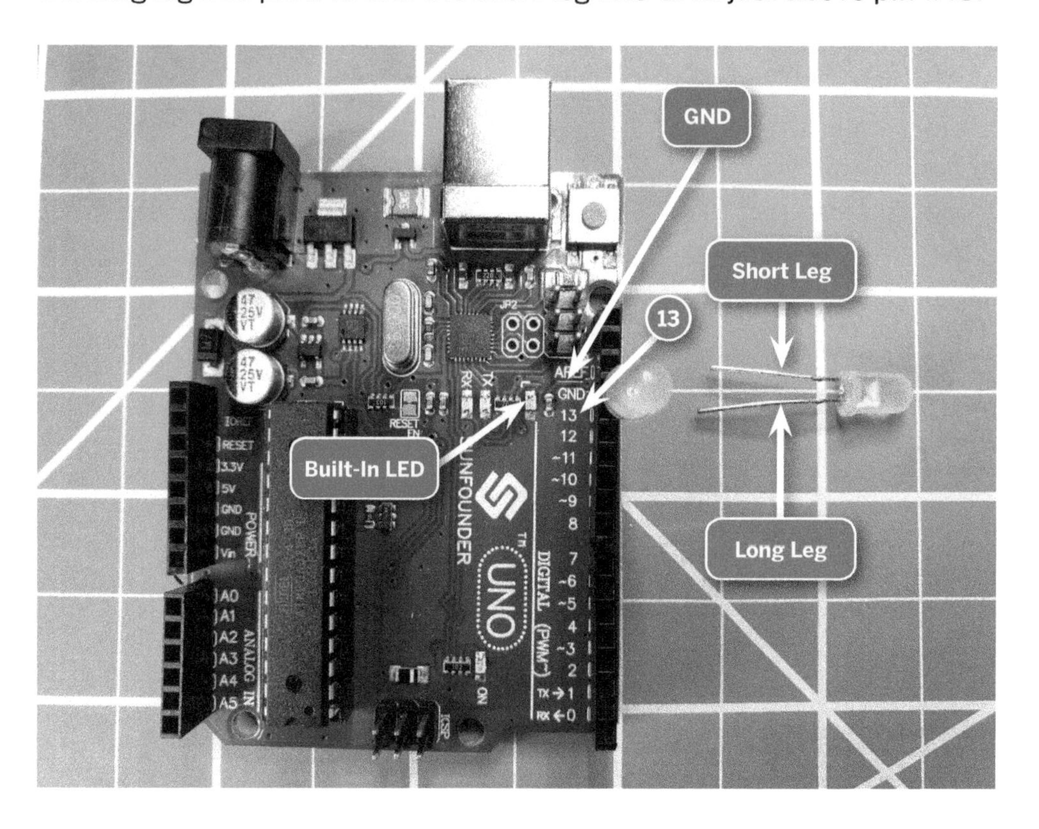

The Most Useful and Amazing Hand Tool

MATERIALS USED IN CHAPTER 2:

- **Crop-A-Dile**
- **$1/3$-inch head-$3/16$-inch center or 4mm Eyelets**
- **Craft Sticks**
- **Paper Straws**
- **4mm Nylon Screws**
- **4mm Nylon Nuts**
- **20- to 30-Gauge Wire Stripper**

My background is as an engineer and a teacher. When I was an engineer, I would often design tools that I thought necessary for the job in front of me. As a teacher, I am always looking for common, inexpensive classroom materials to use in projects. Early in the 2018 school year, we received a donation of 10,000 craft sticks. I wanted to find a way to use the craft sticks as building materials for a variety of robotics projects in the K–8 classroom. I wanted to be able to use the craft sticks to build complex moving mechanisms attached to servos and controlled by microcontrollers such as Arduino. The goal was for students to be able to build a truly useful $5 robot of their own design. I have always wanted my students to use realistic engineering tools, bike screws, nuts, bolts, and cutting tools. Engineers do not use glue and duct tape, so I do not want my students using those either. Just because my students only had craft sticks as a building material does not mean that the design should lack precision.

The primary weakness of craft sticks as building material is their propensity to crack along the wood grain. All craft sticks have the grain of the wood running lengthwise along the stick. This makes the sticks extremely strong in one direction only. As soon as you try to punch a hole in one, it will split. This is exactly the kind of situation where an engineer would spend time designing a special tool for the task at hand. I spent three months trying to design a tool that would punch a hole in a craft stick without cracking it. After three months of frustration I walked away from the project and looked again for tools that might already exist for this task. I discovered that I could utilize a tool used by the scrapbooking community for a different purpose. The tool is called a Crop-A-Dile®.

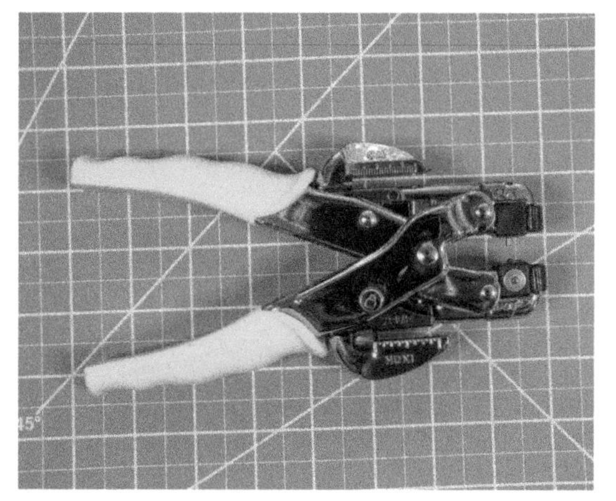

Baby Crop-A-Dile

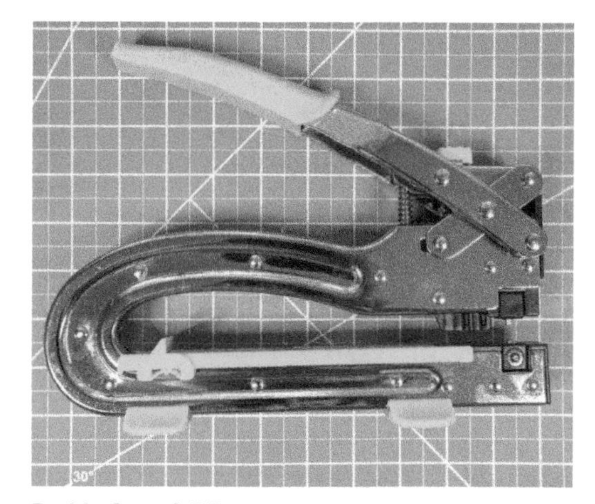

Daddy Crop-A-Dile

There are two versions of the tool, and both versions function exactly the same. The only difference between the tools is the throat depth (that is how far in from the edge of a sheet the rivet can be placed). In the case of the smaller tool (Baby Crop-A-Dile), there is a 2-inch depth restriction from the edge of the sheet; for the larger tool (Daddy Crop-A-Dile), there is a 6-inch limit. This means that with the smaller tool, you are restricted to sheets that are 4 inches across (double the 2-inch throat). You are restricted to 12-inch sheets with the larger tool.

CROP-A-DILE ANATOMY

All of the projects in this book can be constructed with the smaller Crop-A-Dile, so that's the one that we're going to use in this example. The larger version of the Crop-A-Dile is functionally similar. The Crop-A-Dile is asymmetrical, which means that the top and bottom sections perform different functions. Pay close attention to how you hold the tool when using it. The top section of the tool has what I will call a "wiggly tooth." It's easy to identify, just stick your fingers in the jaw of the tool and feel around for the wiggly tooth. This tooth will always go into the top of the grommet (grommets look like mushrooms because the top is the widest part) and center the grommet as it is set into shape. Each "ear" on the tool contains a circular punch of a different diameter ($\frac{1}{8}$ inch and $\frac{3}{16}$ inch).

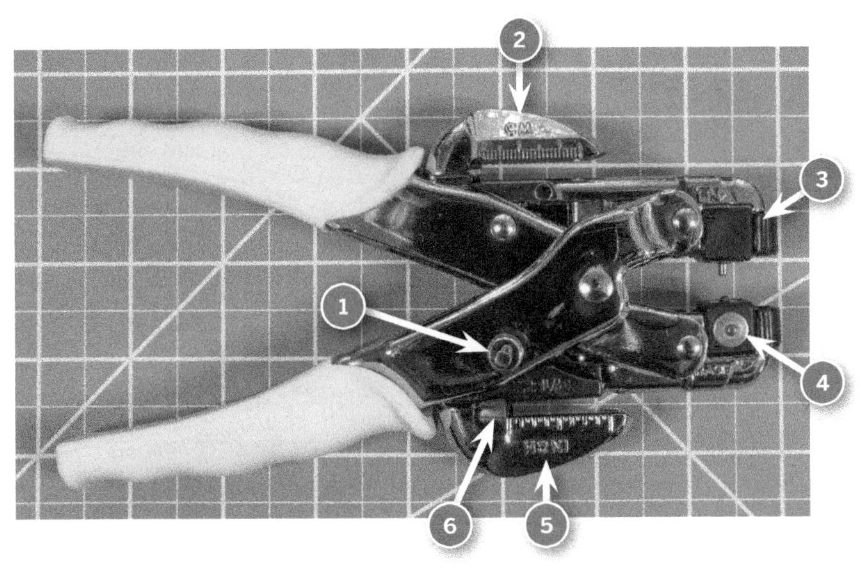

Figure 2.1
Lock button (#1); ⅛-inch punch (#2); Wiggly tooth (#3); Mandrel (#4); $3/16$-inch punch (#5); Adjustable depth gauge (#6)

SETTING UP THE CROP-A-DILE

Grommets come in all different shapes and sizes. It is important to use the correct size wiggly tooth (centering pin) and mandrel for the particular size grommet. We are using grommets with an inner hole diameter of $5/32$ inch (or 4mm). With that size grommet, you should set up the wiggly tooth and mandrel to match pictures below. The photo on the left shows the two pieces rotated to the correct position.

PUNCH 'N SQUISH

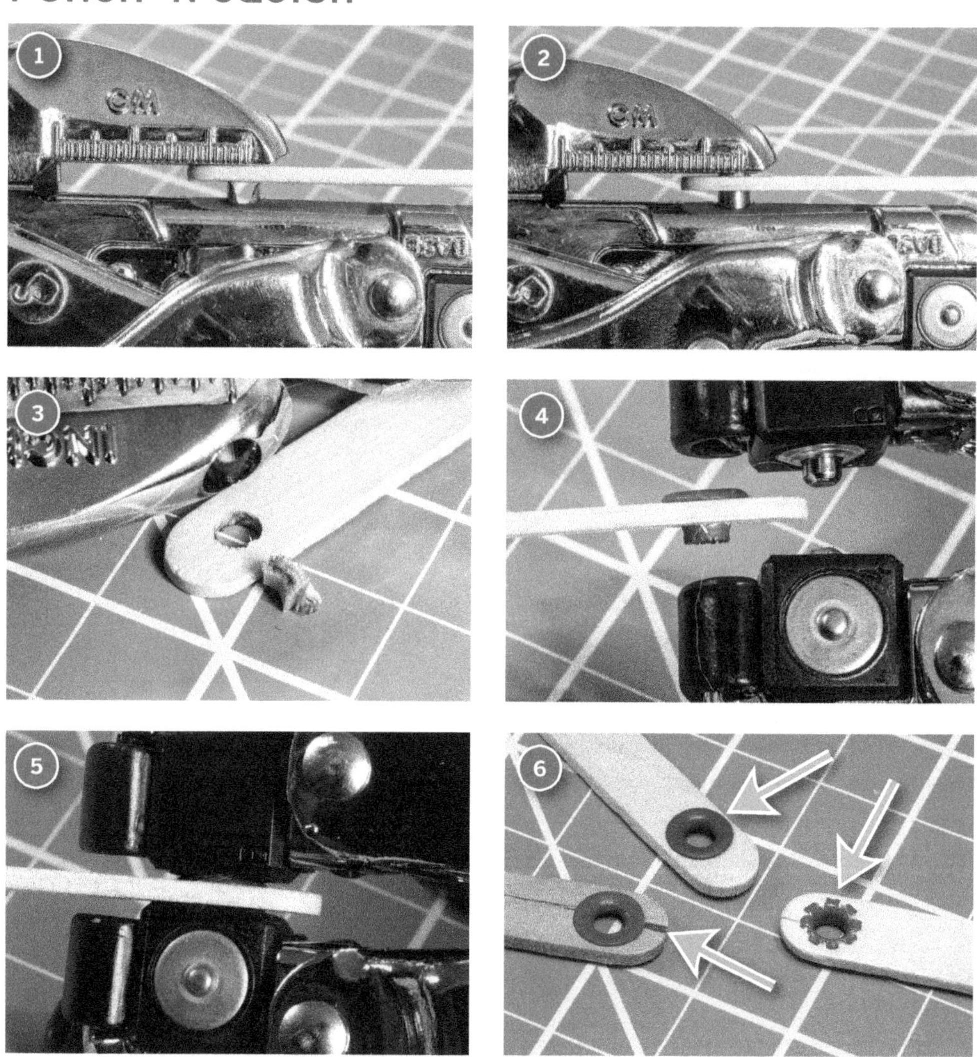

Figure 2.2

Showing side view of the ear with the 3⁄16-inch punch lined up with the end of a craft stick (#1). The 3⁄16-inch punch going through the craft stick as the handle is squeezed (#2). The completed hole punched in the craft stick (#3). The grommet inside the hole in the craft stick with the fat side facing up toward the wiggly tooth (centering pin) (#4). After aligning the grommet face-up inside the jaws, squeeze the jaws to close the grommet between the two surfaces (#5). The grommet should look like this (#6). The one to the left shows how it is possible to crack the craft stick by squeezing too hard. The one on the right shows the underside of the grommet after squishing it.

Note that the wood pushed from the inside of the hole will usually stick inside the tool as opposed to being ejected from the tool. There is no need to clean it out every time as the new wood will push out the old wood with each use.

USING THE CROP-A-DILE

To become more familiar with the Crop-A-Dile tool, we will build the triangle pictured in image 8 on page 29. Triangles are useful, rigid, and strong. The triangle in image 8 has three possible pivot points and can hold a load of 40 pounds vertically! You can build one similar to this example in approximately three minutes simply by referring to page 28. However, I would encourage you to read through all the instructions because I am going to introduce some commonly used engineering terms and concepts around each of the parts within the building process.

FEELING SCREWY

The first component that we want to understand after the Crop-A-Dile is the screw. This might sound simple, but stick with me to make sure that you understand all the terminology and physics.

The screw has:

- threads
- pitch (the distance between the threads)
- a head (the top of the screw where the screwdriver inserts)
- a nut (a corresponding hexagonally shaped nut with matching internal threads)

The most important specification for a screw is the diameter and pitch of the threads. Anytime you are ordering a screw, you need to specify the diameter of the screw and the distance between the threads. The two main formats used are SAE (Society of American Engineers) and metric. SAE is popular in the United States, and the rest of the world uses the metric system. There are, however, differences between the two.

An SAE screw specification might appear as "2-inch 4-40 Pan Cross."

- "2 inches" refers to the length of the screw from just below the head to the tip.
- "4" refers to the diameter of the screw shaft.
- "40" refers to the distance between threads (in this case 40 threads per linear inch).

- "Pan" refers to the cross-sectional shape of the head of the screw.
- "Cross" refers to the shape of the screwdriver tip needed to turn the screw.

A metric specification might be given as "15mm M2 Pan Slotted."

- "15mm" refers to the length of the screw from under the head to the tip.
- "M2" refers to the diameter of the screw.

You may have noticed that in the metric example, the distance between threads (pitch) is not noted. This is an important distinction for people just starting out building machines. Metric thread pitch is standard by default, unless otherwise specified. In this book if the project requires an M4 screw, that is all I will specify.

But what about length? If metric thread pitch is standard unless specified and you can use whatever kind of head style you want, then why not just buy a giant bag of screws and simply cut the length you need each time? That is exactly what we are going to do! Therefore, the screw specifications in this book will be something like this: "a bag of M4 screws as long as possible."

This project will use nylon screws that are soft enough to be cut cleanly with wire cutters using very little effort, yet be strong enough to function as pivots and joints in the machines you are going to build.

ANATOMY OF A SCREW

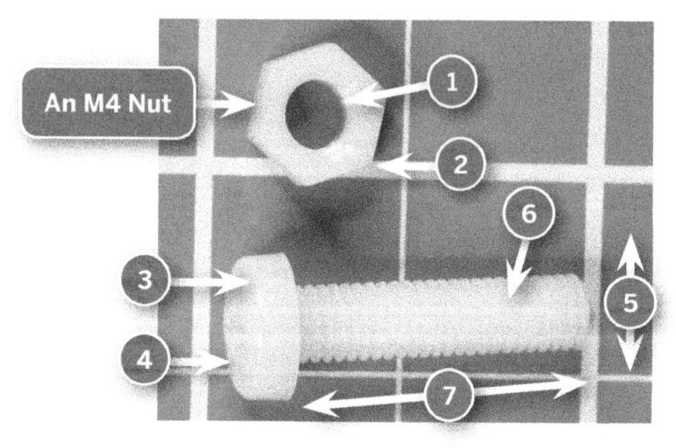

Figure 2.3

Diameter (#1) and (#5); head shape (#2) and (#3) (typical shapes are pan head, round head or flat head); screwdriver bit style (#4) (cross, slotted or hex); pitch (#6); and length from under the head to the tip (#7).

CUTTING SCREWS

One of the tools that we are going to be using frequently for our projects is the wire stripper. A wire stripper is a handheld tool that looks like a pair of scissors but has several circular notches that will allow you to cut through the plastic insulation surrounding a wire without cutting the copper wire itself. We will address the functionality in later chapters. For now, we want to focus on the cutting area, located just before the pivot on the wire strippers, which we will use to shear off the unwanted length of a screw. This is because it is easier to have just one bag of long screws all the same length rather than having 10 different bags of screws of differing lengths, multiplied by several different diameters for each project.

Based on a great deal of personal experience, I can recommend two models of wire strippers, both available on Amazon:

1. Hakko CHP CSP-30-1 Wire Stripper, 30-20 Gauge
2. Eclipse Tools CP-301G Pro's Kit Precision Wire Stripper, 30-20

ANATOMY OF A WIRE STRIPPER

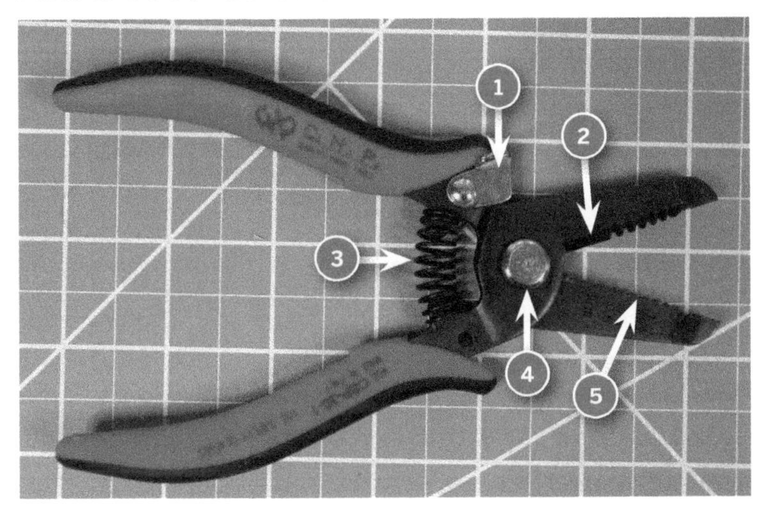

Figure 2.4
Locking tab (to lock closed for safe storage) (#1); cutting section (#2); spring (#3); pivot (#4); circular notches for stripping the insulation from a wire (#5).

For this project you will use the cutting section of the blade.

CONSTRUCTION STEPS

Follow the pictures in below to assemble the triangle. Your overall goal is to place grommets in both ends of one paper straw and in both craft sticks.

1. Push a 4mm nylon screw through the grommet on the end of each craft stick (#1).
2. Twist the nut on each (#2).

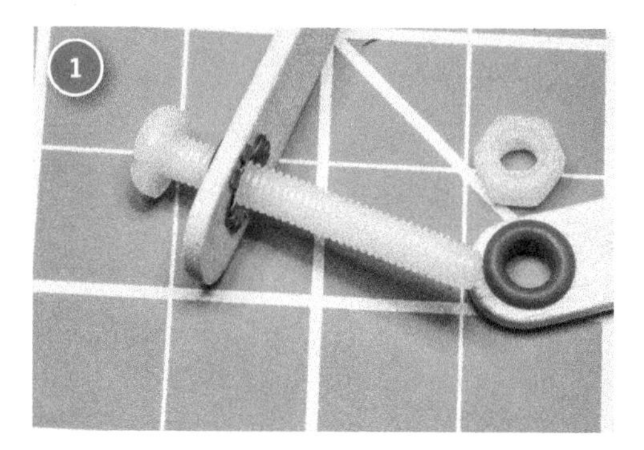

3. Cut the screws to the desired length after adding the nuts (#3, #4, and #5).

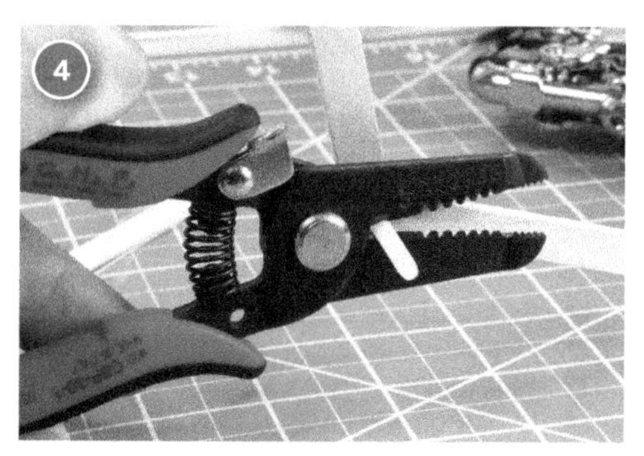

4. Pinch the ends of the paper straw (#6).
5. Punch a hole in the straw and add grommets to each end (#7).
6. The finished assembly (#8).

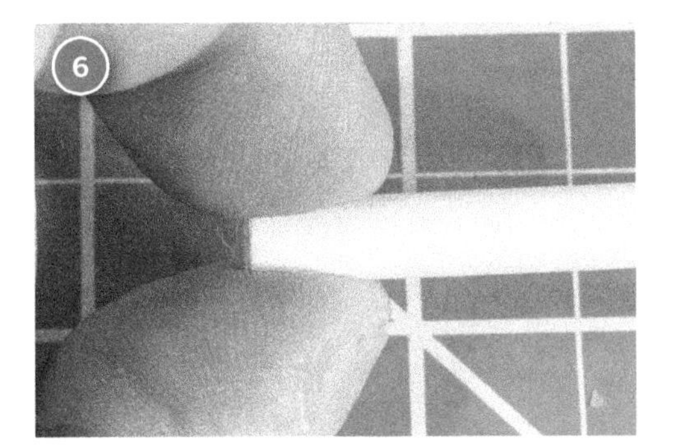

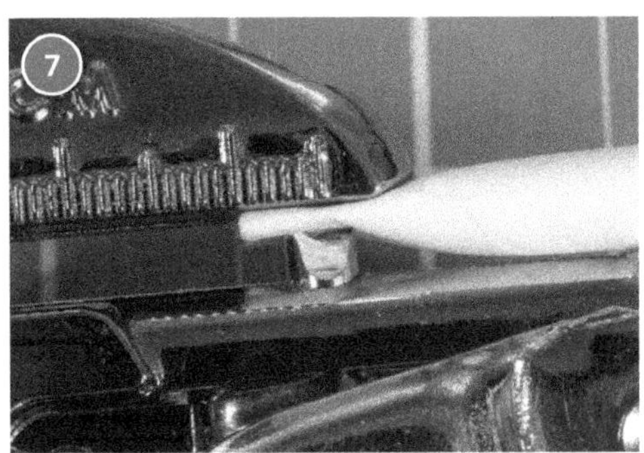

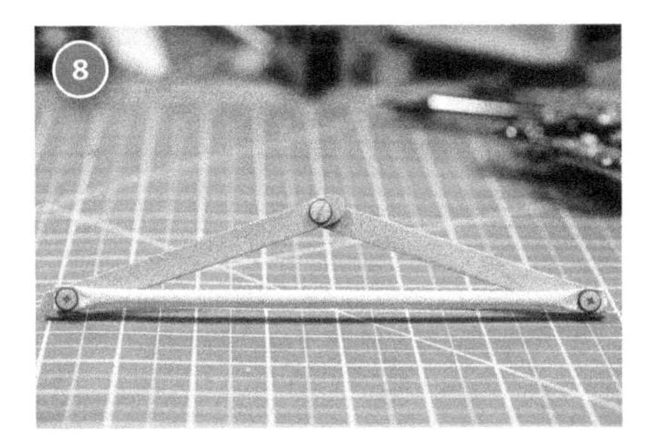

After completion of these steps, stop to think about what you have accomplished by building a simple triangle. Being able to do so allows us to build objects that rotate on a single axis, such as a propeller, or build rigid structures, such as the body of a robot. Because the materials are inexpensive, you now have the capacity to create your prototypes quickly and recycle the parts that do not work out. This is tinkering in its truest form. Very often the form of a project will come out of this tinkering process by just playing with the tools and materials you have available to you.

Before going further, I would encourage you to see what you can build by placing the grommet in the center of an item or at the ends of craft sticks. For instance, you might be able to construct something like a clock where you have a minute and a second hand sharing the same rotational axis. These explorations are what you will use to flesh out the final form of projects presented in this book. My goal is for you not to simply copy the steps in these chapters, but actually take a meandering path as you tinker with each new tool, skill, and material presented to you.

IN THIS CHAPTER YOU:

- used the Crop-A-Dile hand tool—the most useful hand tool on the planet for makers.
- squished an eyelet so as to set it without cracking a craft stick.
- learned about the anatomy of a screw and nut.
- trimmed a nylon screw to size.

TAKING IT FURTHER

There is a wide variety of eyelets out there and an even wider variety of materials to punch and set eyelets into. For eyelets, do an Internet search for "We R Memory Keepers 41595," which will pull up a variety of enameled eyelets with wide heads on them. For materials, consider leather, ⅛-inch plywood, draft board, or even thin sheets of aluminum.

Blink Writ Large

MATERIALS USED IN CHAPTER 3:

- **IoT Relay by Digital Loggers**
- **Male-to-Male Breadboard Jumper Wires**
- **Arduino Uno R3**
- **Small Slotted Screwdriver**

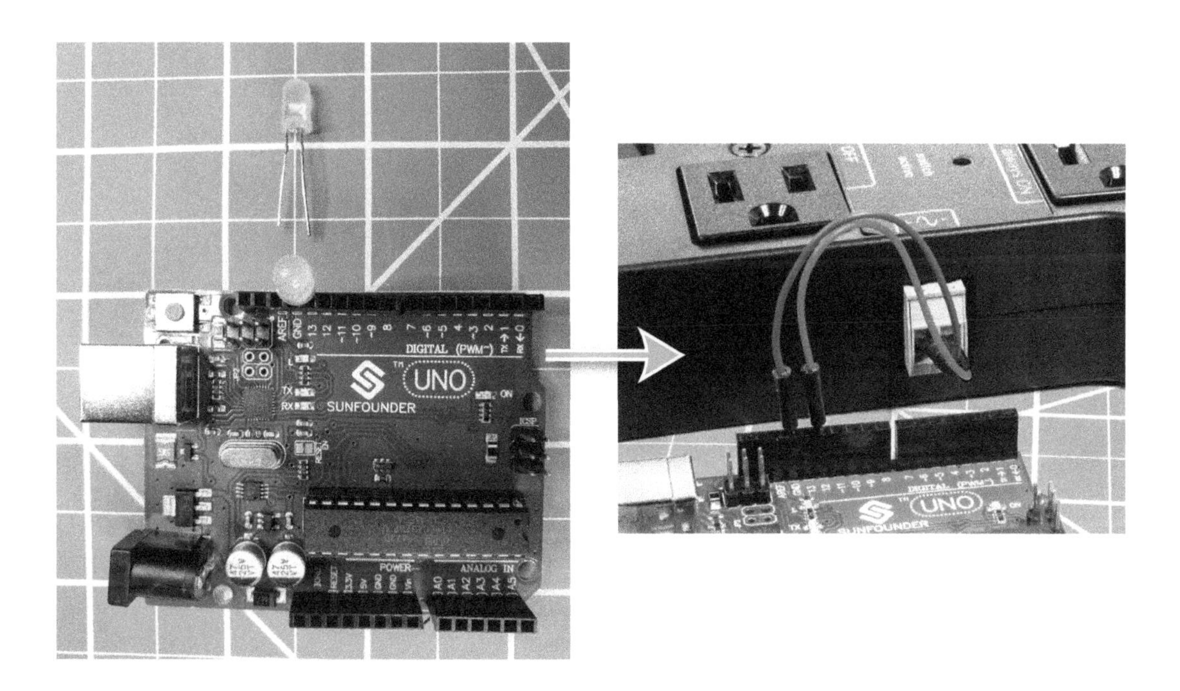

In chapter 1 we set the groundwork for working with the Arduino. We used an example program called "Blink." This program demonstrated what a loop is and how to use it to control the timing of an LED blinking on and off. For the project in this chapter, we will use the same sample program to blink a very large light. We will be able to control a table lamp or anything that you can plug into a standard wall outlet. We are going to use this example program in much the same way your fingers turn on and off a wall-mounted light switch.

How can we control so much electricity with such a small board? We can do this by taking a very small amount of electricity that is switched on and off inside the Arduino and use that to signal a much larger switch that actually controls the flow of electricity on a much larger circuit. You can think about it like this: When you walk into a room and use your finger to flick a light switch on the wall, you use a small amount of energy and the muscles in your fingers to move the switch from the on position to the off position. The switch itself only needs the energy your finger is putting on it to toggle on and off as it controls a much larger amount of electricity to power the lights in the room.

If you did not complete the Taking it Further section of chapter 1, do that now. In that section we added an external LED to blink along with the built-in LED on the Arduino. What we are doing in this chapter is removing the LED and using that electricity to send a signal to a much larger switch that will control something like a table lamp that we would normally plug into the wall. We are literally unplugging the external LED and connecting its signal wires to a bigger switch.

A relay is a switch controlled by a small amount of electricity, instead of one that you flip on and off with your finger. In this case our switch is called an AC/DC control relay. What this means in plain English is that we are using the very small amount of DC (Direct Current) that was powering an LED to instead send a signal to a relay controlling AC (Alternating Current) which is the kind of electricity that flows inside your house. Specifically, we will be sending a 5-volt DC signal to control the 115-volt AC outlet that you can plug a desk lamp into.

The box I recommend using for this is manufactured by Digital Loggers (IOTrelay.com). While this is a very safe project, you should be cautious when working with electricity. As you follow these steps, keep the relay box unplugged from the wall outlet.

SETTING UP OUR RELAY

Page 33 shows the materials needed for this project.

Jumper wires are wires that already have metal ends on them. A quick search of Amazon under "Arduino jumper wires" will provide the user with many different choices. Wires with pointy ends are referred to as male and wires with holes on the end are referred to as female. Wires come in M/M, M/F and F/F versions of varying lengths. I would recommend getting a pack of 40 of each of the three versions.

Pictured below you can see the main power switch for the relay (#1); two outlets "normally off" (#2); one outlet that is "normally on" (#3), one outlet that is always on (#4); the input for our signal from the Arduino (#5); and the male-to-male jumper wires (#6). The outlets that are labeled "normally off" will not have power unless the Arduino is currently sending a signal to the relay. The outlets that are labeled "normally on" will have power when the Arduino is not sending a signal and turn off only when the Arduino is sending a signal. You can use this functionality to turn off the light or turn on a heater in your room with a single signal from the Arduino.

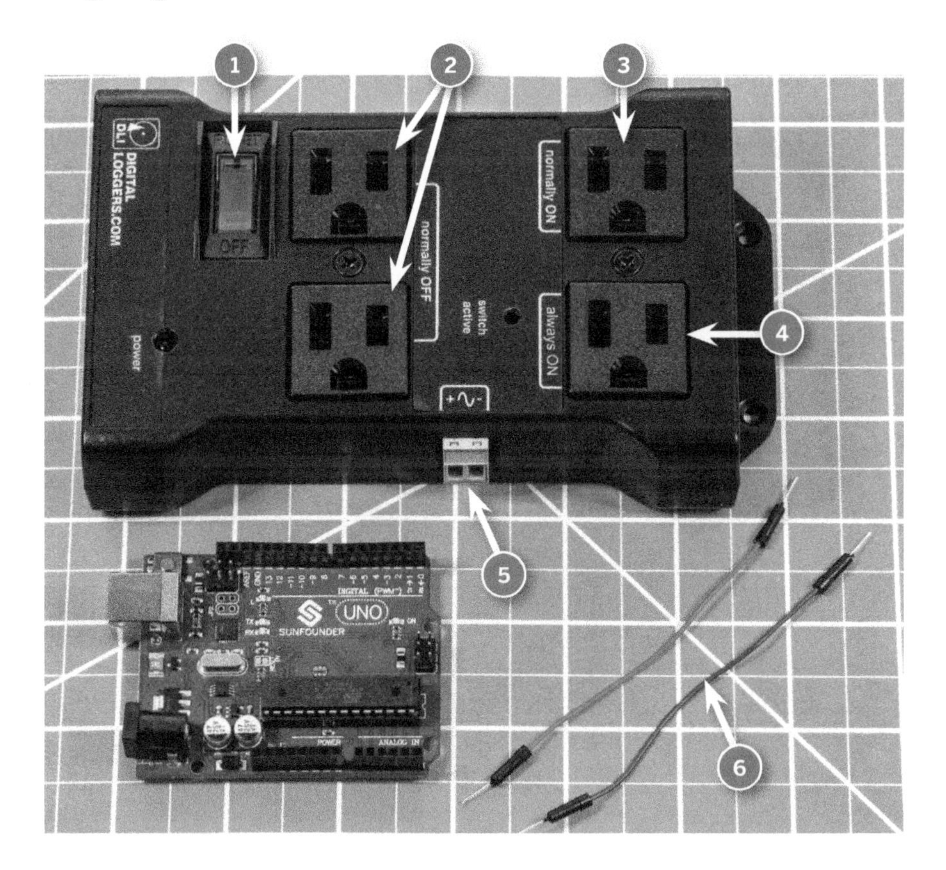

When you are building this project, you are picking up the powerful skill of controlling the flow of electricity to household appliances with the same simple lines of code that we started to explore in chapter 1.

WIRE THE RELAY TO THE ARDUINO

1. Pull the green plug straight out of the side of the black relay box. This green plug is called a "terminal block." Note the two small screw holes on top of the plug.
2. Use a small screwdriver to open each of the two wire receptacles. You should be able to see the metal jaws inside the receptacle opening and closing as you turn the screws clockwise or counter-clockwise. The photo labeled #2 shows one receptacle fully open and one receptacle fully closed.

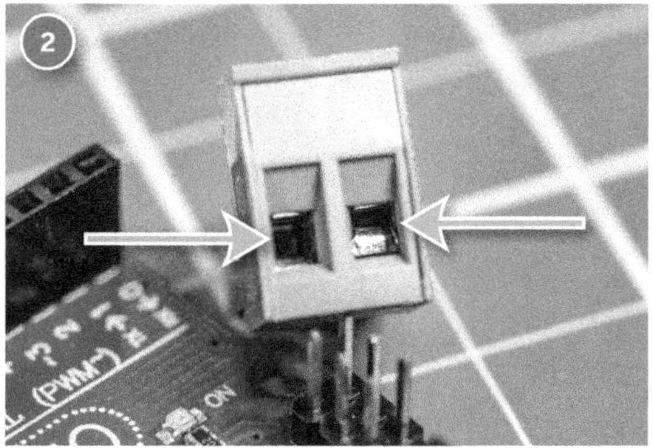

3. In this case we will want electricity to flow from pin #12 on the Arduino into the positive (+) on the left side of the plug, and the electricity to flow back out of the relay on the negative (-) on the right side of the plug and into the Arduino ground (GND) pin. Photos #3 and #4 show the wires inserted into the plug and into the Arduino correctly.

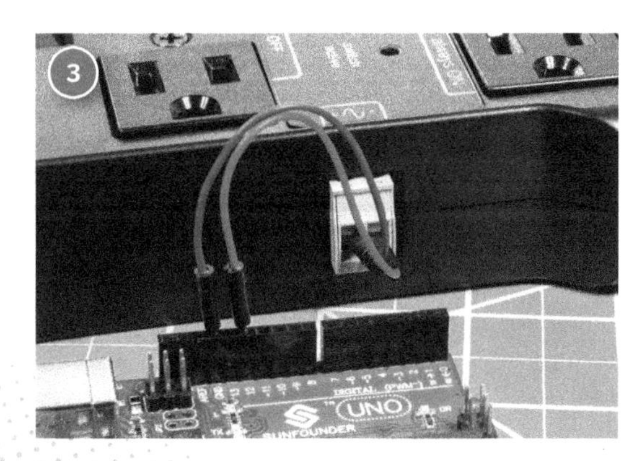

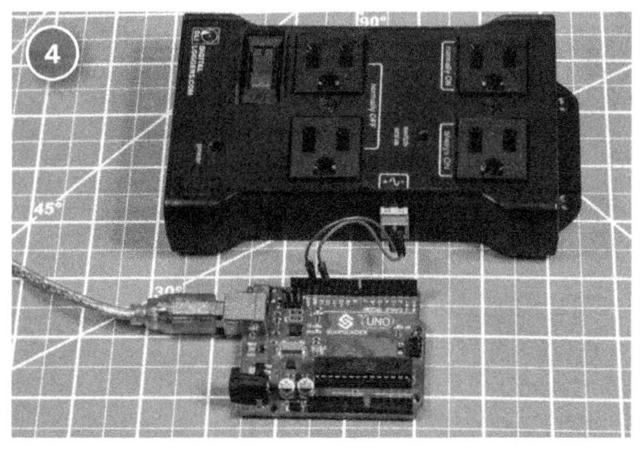

4. Once you have this assembled, run the "blink" program with the relay unplugged from the wall to make sure you have everything set up the same as the example in chapter 1. The only difference will be that you now have the relay box wired in place of the LED.

Once you are comfortable sending a signal to the relay, simply plug the relay into the wall socket, and you are ready to control any household appliance with a few lines of code!

DECODING THE CODE

If you are already feeling comfortable with how the code on page 38 works, you can skip this page. If not, I want to take some time to dissect the code in a little more detail, so you feel comfortable modifying it and exploring variations. Having a deep understanding of how the code works is the first step to writing your own program.

I want to explore the code by stating it in plain English. Read aloud and repeat the following several times:

"I am declaring a new variable that takes the form of an integer, naming it ledPin, and setting the value of this variable to 12. I am then specifying the purpose of the physical pin associated with that variable to be one of an output, not an input. I am now going to repeat in a loop over and over, 'I command that we turn on pin number 12 for 1000 milliseconds (one second).' Then I will send a command to turn off pin number 12 for the same amount of time before starting the process over again."

When I was first learning to code, I was perplexed with why it was necessary to go through the seemingly extra steps of creating a variable, then assigning a value to the variable, and then using that value to specify the pin number on the Arduino that we wanted to turn on or off.

It turns out the reason we are taking those few extra steps is because we have the ability to do many things every time the code goes through another cycle in the loop. For example: We could ask that the value of that variable increase by one every time we go through that loop. If we put an LED on every single pin from 1 to 13 and started the value of the variable at one, then we could turn each of the 13 lights on in order. This knowledge will become important in our future projects.

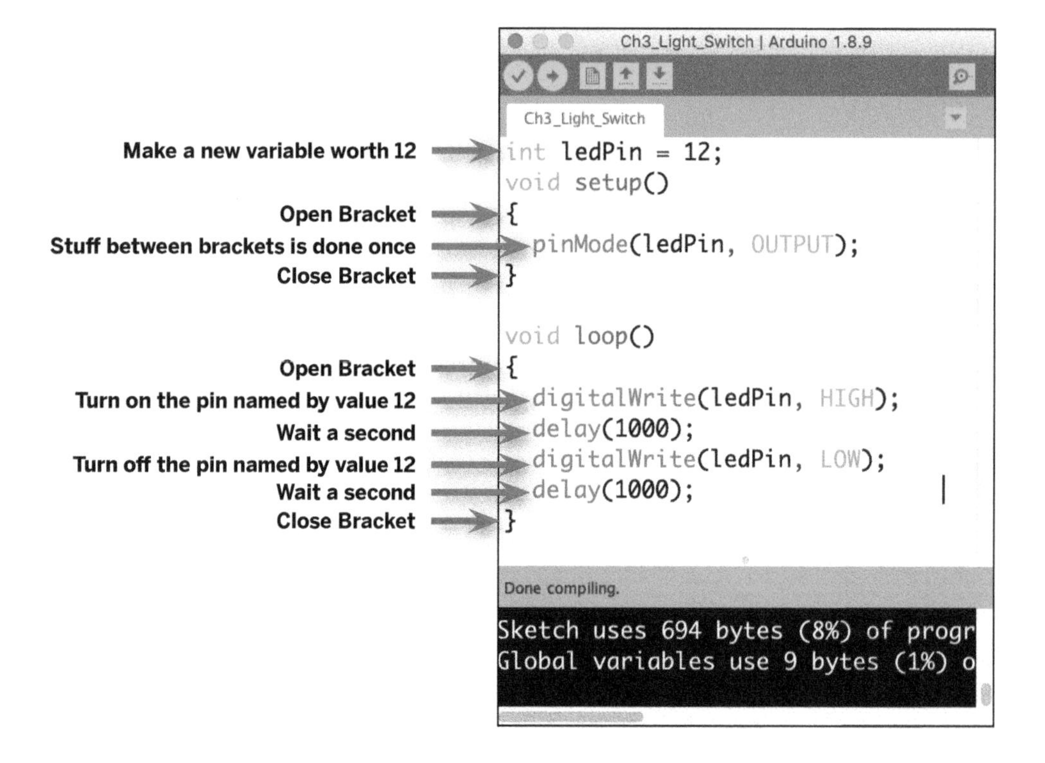

- worked safely with a lot of power.
- used a small signal on the Arduino to control lights that you normally plug into the wall outlet in your home.
- used jumper wires.
- used a terminal block.

TAKING IT FURTHER

Now that you are capable of switching 110 volts AC on and off, you can think about other relays in smaller and more open packages. Search the Internet for Arduino relay, and you will see a variety of options you can use to get much smaller versions of this large relay to integrate into projects. As you venture beyond this book, keep in mind that the IoT relay we used in this chapter was set up for safety and that you should read more about working safely with 110-volt AC circuits before doing more work in the world of relays.

4 Flag Waver

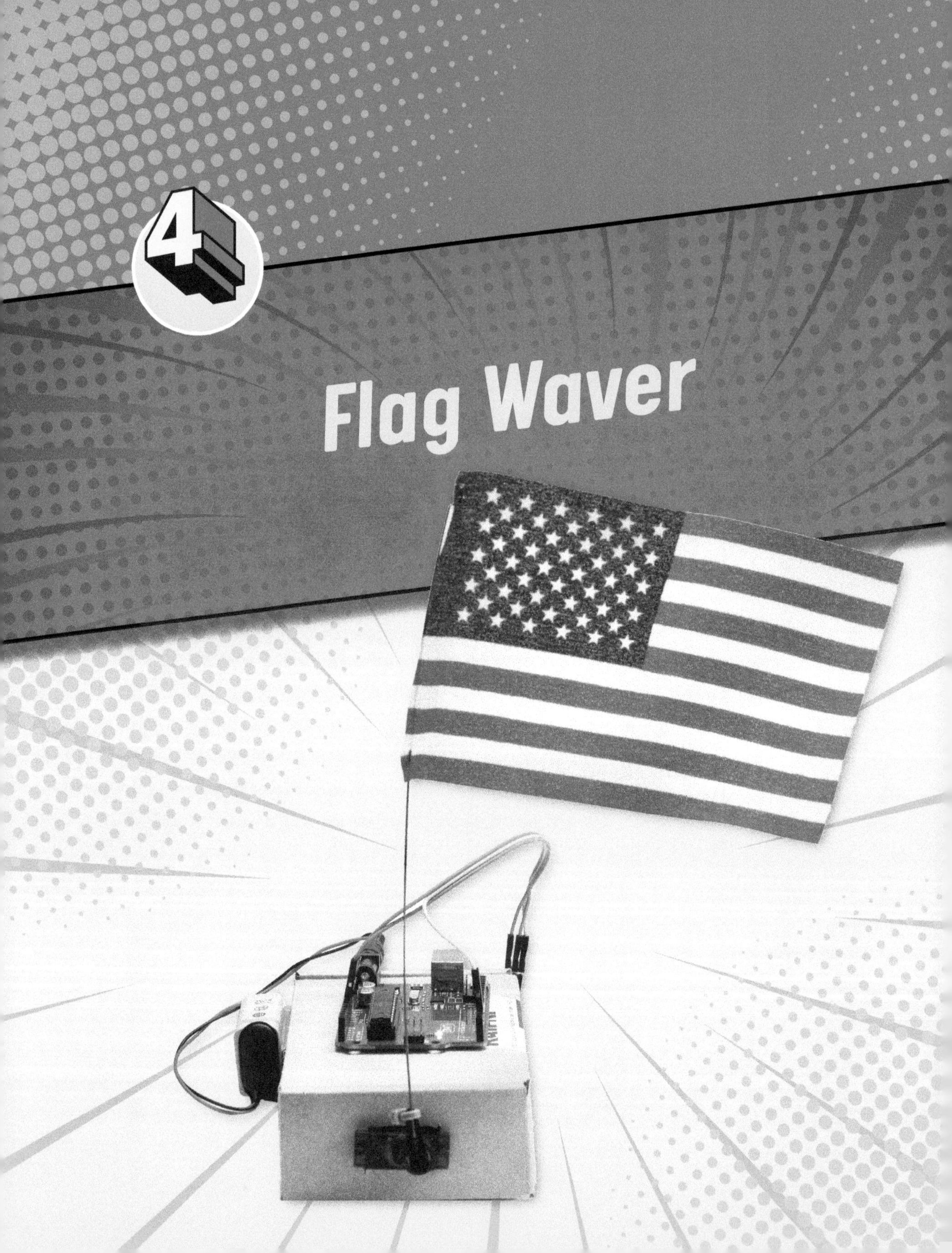

MATERIALS USED IN CHAPTER 4:

- SG90 Metal Gear Hobby Servo
- Male-to-Male Jumper Wires
- Small Phillips or Cross-pattern Screwdriver
- Small Corrugated Cardboard Box
- Hobby Knife
- Fine-Point Permanent Marker
- Aduino R3
- 9-volt Battery
- 9-volt Battery Clip to 2.1mm x 5.5mm Barrel Jack Adapter
- Jeweler's Conical Jaw Pliers
- 0.039" or 0.032" Music Wire
- Small Zip Ties

In chapter 3 we learned to control large amounts of electricity with a small signal from the Arduino. In this chapter we are going to cause something to physically move. The mechanism that we are going to build is suitable to add motion to holiday window decorations or to wave a small flag. You will need two cardboard boxes at least an inch tall and at least the size of a bar of soap along each side. We are going to use this cardboard box as a physical body, or chassis, for the machine. Up until now, we have wired Arduino to another component without integrating all of the components onto one body.

WELCOME TO THE THREE-WIRE CLUB

In chapter 1 you learned how to get started with Arduino in both a virtual environment like Tinkercad as well as programming the physical board. In chapter 2 you learned how to use the Crop-A-Dile hand tool. In chapter 3 you took the original code from chapter 1 and used it to turn a desk lamp on and off. In this chapter we are going to explore a concept that separates engineers from non-engineers. This is where most people new to the subject tend to fall away or stop understanding the fundamental level of how things work. My goal in this chapter is to get you comfortable with something called the third wire.

When you ask most people how many terminals a battery has, they respond with the correct answer "Two: One positive (+), and one negative (-)." Behind that answer is an understanding that electricity will flow from the positive (+) end of the battery, do some work (like making a motor move or turn on a light bulb), and then somehow return to the negative (-) terminal on the battery to complete the circuit. When we triggered the relay box in the last chapter, this is exactly what happened. Two wires, one for the electricity to go out, and one for the electricity to come back in, sent a signal to the relay box which switched the larger circuit on or off. However, almost every sensor and every output device we can imagine (other than simple lights and motors) have at least three wires, if not more, that need to be hooked up correctly in order for them to work. That is the dividing line between people who are intimidated by engineering and those who start to feel comfortable with the naked circuit board.

Pictured on page 43 is a special type of motor called a servo. You will note that the servo has three wires coming out of it. A servo is a motor that can not only turn in both forward and reverse, but also a specific number of degrees and hold a position. We are going to have the Arduino command the servo to first move to 50 degrees and hold for one-tenth of a second, and then move to 130 degrees and hold for one-tenth of a second, and to continue that loop indefinitely. This is a little different from the LED commands to turn on and off, that were introduced in chapter 1. In this case, we are not only telling the servo motor to move, but to move to a specific position at a specific time and then move to another position at another time. In the case of the LED, it was a binary choice: We could either send electricity to the LED and it would light up, or we would not send electricity to the LED and it would not light up.

In order to control something like this servo, we will need the third wire to send data to the servo motor so that it can subsequently interpret the data and move to a specific position. If the signal traveling along the third wire is also comprised of electricity, then how does it get back to the Arduino without its negative (-) wire? The simple answer is that negative (-) wires can be shared when they are going to the ground (GND) pin on a circuit. In this case, we will have a positive wire carrying 5-volt DC, a negative wire to complete the circuit and provide the power for the servo motor to move, and a third wire to carry signal data to the servo motor. However, the third wire will share the "ride home" on the same negative (-) wire so we will only have a total of three wires.

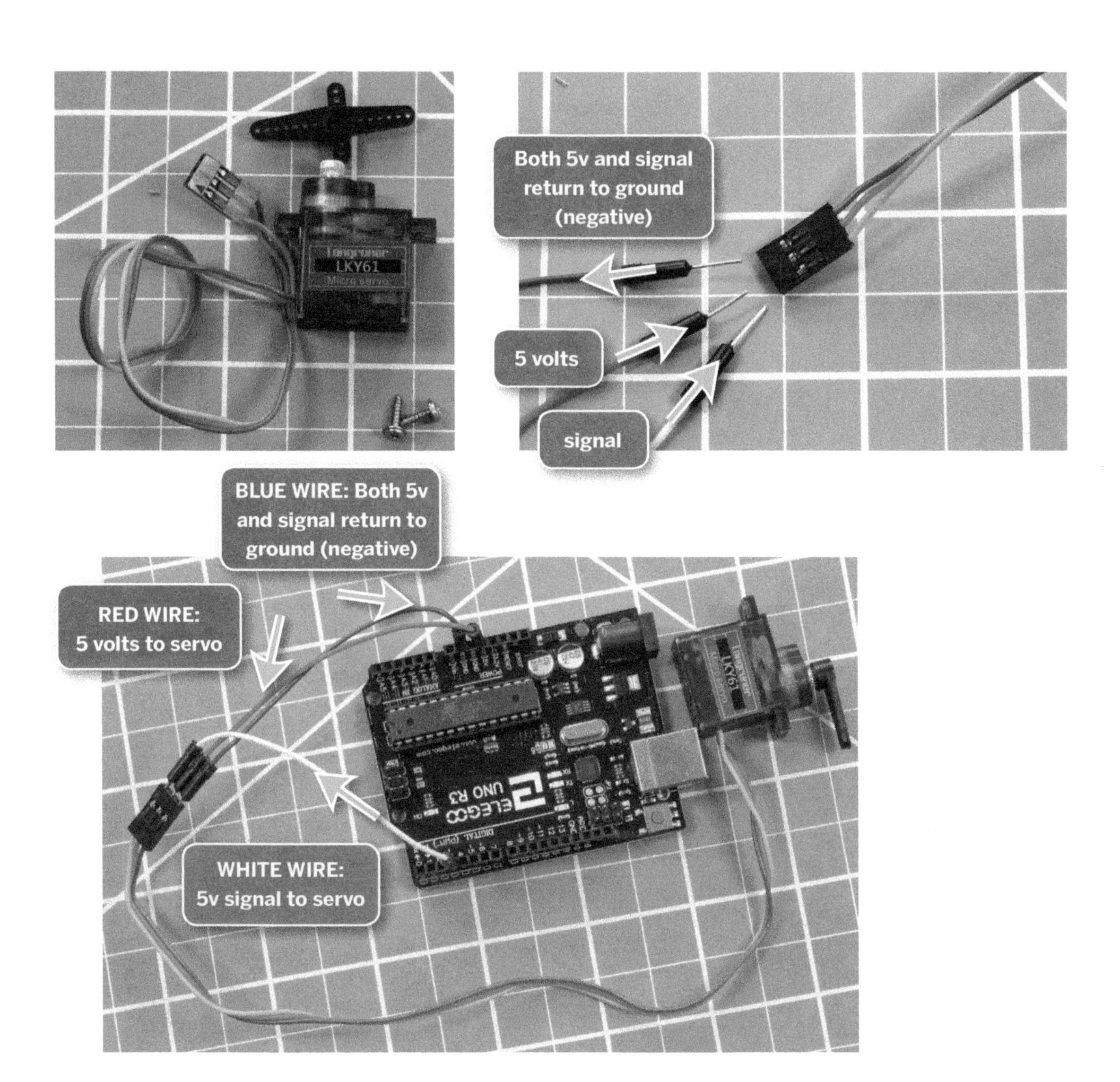

In the picture above you can see the full circuit laid out with all three wires: power (red), signal (white), and shared ground/negative (blue). As long as you are comfortable with the concept of power and signal electricity flowing back from the servo to the Arduino through a shared ground/negative wire, and thus forming a complete circuit, then we should be ready to move on to more complex circuits. If this makes sense, take pride in the fact that you are the newest member of the three-wire club! The vast majority of sensors and output devices you might want to use in future projects operate on the same concept and are set up in a similar fashion.

To give you a challenge, I have plugged the signal wire into pin #3, instead of pin #12 as it was in chapter 3. Take a moment to review the code from chapter 3 and think how you might modify the code to turn on the LED from a different pin but keeping the same ground. The code appears below.

Ch3_Light_Switch | Arduino 1.8.9

Ch3_Light_Switch

```
int ledPin = 12;
void setup()
{
  pinMode(ledPin, OUTPUT);
}

void loop()
{
  digitalWrite(ledPin, HIGH);
  delay(1000);
  digitalWrite(ledPin, LOW);
  delay(1000);
}
```

Done compiling.

```
Sketch uses 694 bytes (8%) of progr
Global variables use 9 bytes (1%) o
```

What number would you change in this code to move the output from pin #12 to pin #3?

ASSEMBLE THE SERVO

Servos are rated in torque. Torque is the amount of twisting force that a motor can produce. This is separate from revolutions per minute (RPM). To arrive at a total horsepower rating for any motor, you take the torque rating and multiply it by the RPM at which the motor is spinning. There is no need to go into this more deeply other than to say torque is an important rating for servo motors, and they are sold with the torque specification as the primary label. The most common servo for small-scale hobby and engineering projects are the 9G or

MG90S. These servos come with either plastic or metal gears. In the past, whenever buying in bulk, I would buy ones with plastic gears because they cost less. However, their failure rate is quite high. If this is your first time delving into a project that uses the servo motor, I would do an Internet search for MG90S Metal Gear Servo and buy a pack of metal-geared servos. These should come with a torque rating of Stall Torque: 2.0kg/cm (4.8V). This rating simply means the servo can produce 2 kg twisting force when the lever on the shaft is 1cm long, and it will take 4.8 volts to make that happen. If you want to ponder this a little bit, think about the 2.0 kg rating as the strength of your shoulder muscle and the cm as the distance from your shoulder to the weight that you are trying to lift. If you are holding 2 kg of weight in your hands with the arm fully extended away from your body and try to move that weight up and down, you will experience more effort than if you held the weight close to your body and tried moving it up and down vertically the same distance.

Servos come with a few pieces of plastic that have small holes in them. Note that one might be shaped like a "T" or a cross, another might have symmetry across a large hole in the middle, and another might have a single arm coming out of the middle hole. In each case the larger hole is meant to fit over the metallic shaft on the servo and be screwed down to hold it in place while the smaller holes serve as attachment points for the mechanism/arm/lever used. These pieces of plastic are called "servo horns." Servos of every size come with a variety of servo horns to maximize suitability for a variety of projects. For this project, select the smallest servo horn with the single arm and place it, without screwing it down, onto the servo.

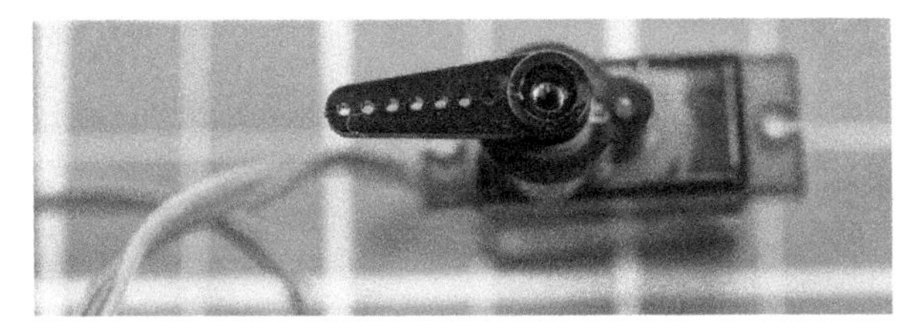

Our servo moves in a 180-degree arc. With the servo unplugged and a single servo horn temporarily attached, rotate the arm *slowly* from side to side. No matter where you attach the servo horn onto the shaft, it will only rotate 180 degrees from left to right. The servo does not know in which position you placed the servo horn column. This 180-degree arc is simply a mechanical limit on the servo and will not change no matter where you place the servo horn on the shaft. As you see below, place the servo horn on the shaft such that the servo horn is in the middle of the 180-degree travel arc. Once you have found this position, use the screw that came with the servo horn to attach it to the servo. When setting the servo arm, if you are within 10 to 15 degrees, you are doing well.

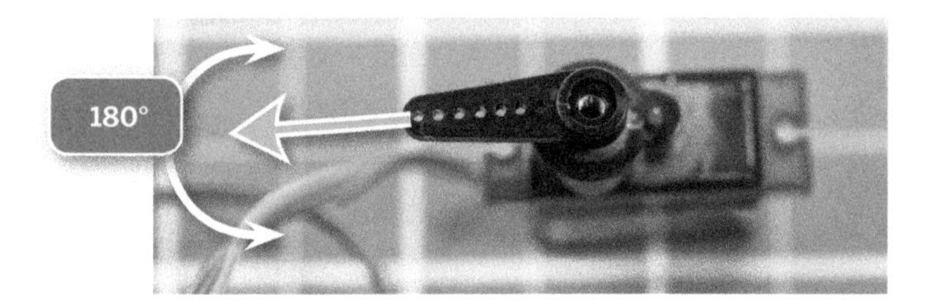

LOVELY CARDBOARD

Many engineers cite CAD (computer-aided design) as the most useful prototyping tool to which they have access. These programs allow engineers to draw three-dimensional representations of machine parts before they ever build a prototype. You already have access to a simple version of CAD if you poke around further in the Tinkercad interface. However, during my time as an engineer, I found a different form of CAD that is much more useful to iterate ideas quickly. My twist on the CAD acronym is cardboard-aided design. I really enjoy grabbing a random cardboard box, a marker, and a hobby knife and starting to prototype directly in a physical space, bypassing the traditional CAD until I have a better idea of how my machine will work.

Follow these steps to mount the servo on the cardboard box:

1. Use a fine-point permanent marker to outline the bottom of the servo on the side of the box. The side of the box will be parallel to the window and perpendicular to the floor when we are done. The servo's 180-degree arc of motion will be waving the flag or moving the decoration up and down in relationship to the floor.
2. Use a hobby knife to cut out a hole just a bit larger than the bottom of the servo.

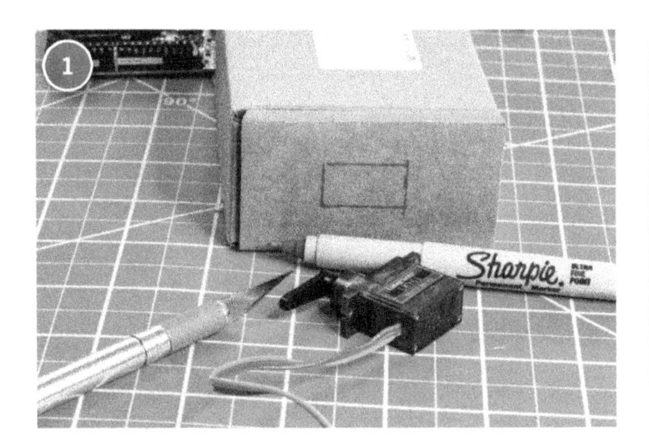

3. If the box has an inner tab, you might need to remove it so you will be able to close the box again. Also, be sure to run the wires back outside of the box unless you want to mount the Arduino inside of the box.
4. Use the self-tapping screws to secure the servo to the side of the box.

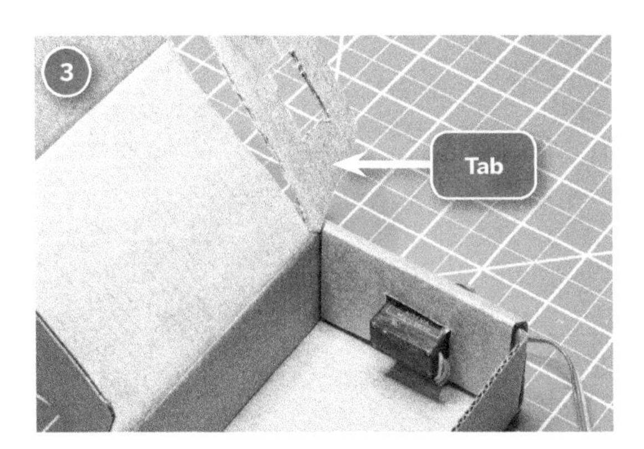

The next step is to mount the Arduino onto the box using the tools and techniques we learned in chapter 2.

ADDING COMPONENTS TO A BODY

We will use grommets and nylon screws to attach our Arduino board to the cardboard box. For ease of assembly in this example, I mounted the Arduino externally on the box. You might choose to mount the Arduino on the inside of the box for a cleaner look. In both cases the process is the same. In the picture below, note the mounting holes on the Arduino board. On page 49, note the final mounting position of the Arduino on the box lid.

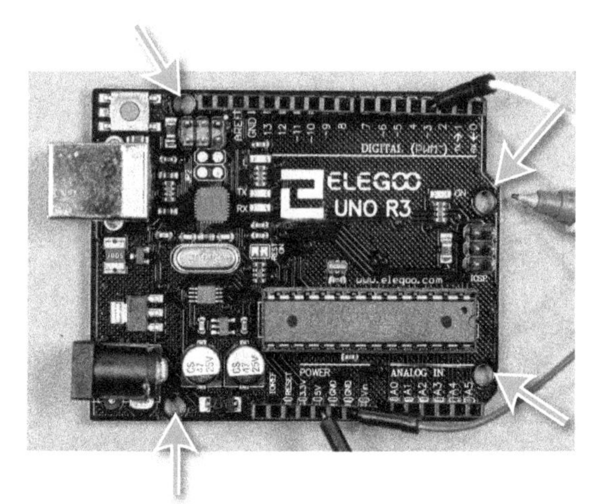

MOUNTING THE ARDUINO ON THE BOX

Follow these steps to mount the Arduino onto the cardboard box.

1. Use a permanent marker to transfer the location of the holes from the Arduino board onto the cardboard.
2. Use a Crop-A-Dile to punch mounting holes into the lid of the cardboard box. In this case I ended up using the larger version of the Crop-A-Dile. Which version of Crop-A-Dile you use will depend on the size of the box you choose and in what location you are mounting the Arduino.
3. Install the grommets with the same method and tool settings that you used in chapter 2.
4. Picture 4 on page 49 is a view of the interior of the cardboard box with the grommets installed.

5. Use 4mm nylon screws, cut to length, and nuts on the reverse side to mount the Arduino board to the box. You will notice that the same red, blue, and white wires are visible from earlier in the chapter when we discussed the three-wire club.

EXTERNAL POWER AND SETUP

Until now we have powered the Arduino board with a USB cable attached to the computer. As we start to make more complex machines, keeping the Arduino tethered to the computer becomes impractical. External power means powering the Arduino with something other than the USB cable connected to the computer. Once the Arduino has been programmed via the USB cable, you can remove the USB cable from the Arduino and apply external power from a battery or power supply. The barrel connector on the Arduino is female and accepts a 2.1mm x 5.5mm male plug. You can use any power supply from 7 volts to 12 volts. Generally, any power supply or battery labeled at 1.5A or 2A is sufficient (as long as the voltage is within 7–12 volts). You can use a larger power supply without danger to the Arduino, though I would stay away from anything with more than a few amps' capacity for your own safety until you know more about power supplies. The worst (and cheapest) battery you can use for this purpose is a 9-volt battery. The reason this is a bad choice is that it has so little electrical storage capacity and will soon die. Even so, I recommend an Internet search for "2.1x5.5mm Male DC Power Plug to 9V" and spending a couple of dollars to buy an adapter cord and a few 9-volt batteries, and then mounting the battery with double-sided tape. I would start here because this is the simplest and cheapest way to get started with external power. In a later chapter we will cover better power supply options, but for now I want to get you familiar with the simplest possible solution.

Note that there is no on-and-off switch in this setup. You simply plug or unplug the battery cable from the Arduino to turn it on or off. One of the beautiful things about the simplicity of the Arduino is it takes almost no time to boot up and start running the code that you have programmed into it. No matter how many times or with what frequency you yank the power cable in and out of the board, it will continue to work as intended. It is a robust, beautiful, and simple machine.

USING THE SERVO HORN

For this project we want to attach a long section of semiflexible wire to the servo horn. To do this, use a pair of jewelers' pliers with conical jaws to bend a 12-inch length of piano wire into the S-shape shown on page 52. I have found the Anezus 4pcs jewelry pliers tool set to be a great value. The best wire to use would be a section of .032″ steel piano wire. Be sure to use a "hard wire cutter" instead of a regular wire cutter. (The regular wire cutter might work, but the tool will become dull after a few cuts.)

1. Create two bends in opposite directions as close to the end of the wire as possible.
2. Note the finished shape.
3. Insert the wire as shown into the hole closed to the center mounting point of the servo horn.
4. Use a zip tie at the top of the servo horn to secure the wire in place.

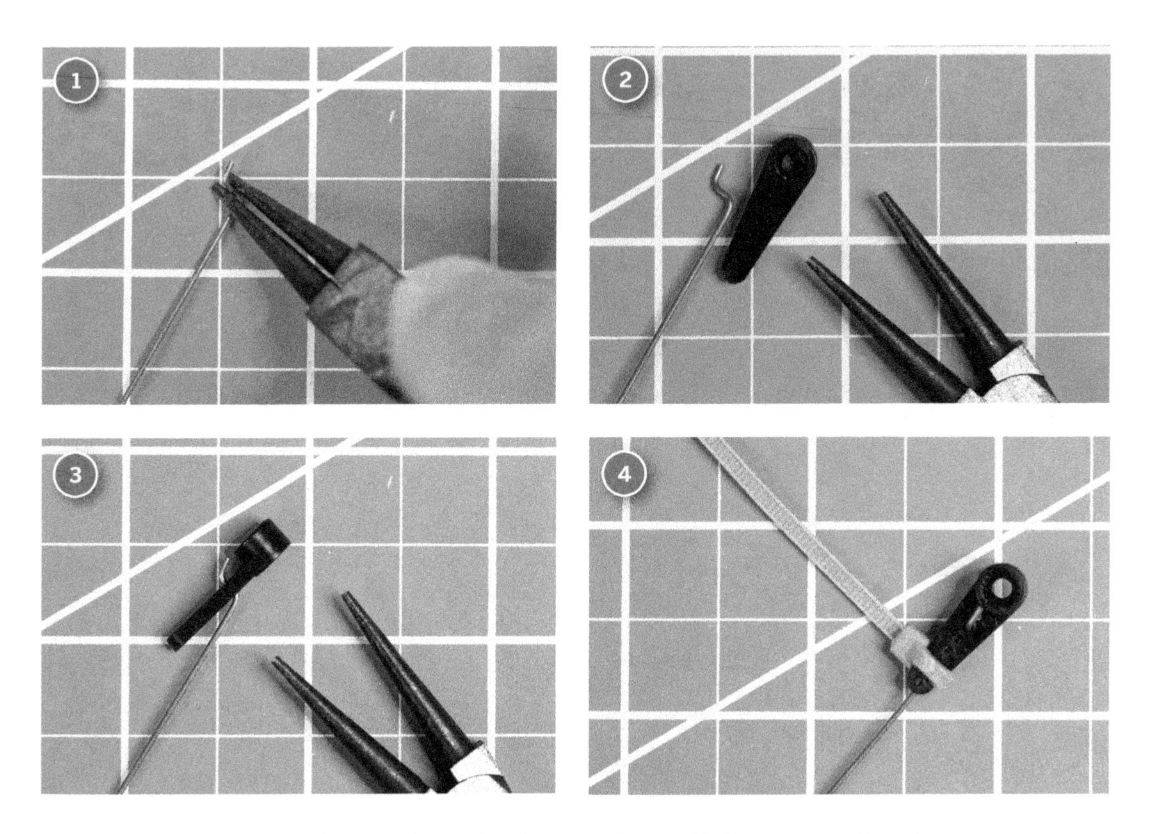

The goal here is to have a length of wire that will flex several inches in each direction on the opposite end from where you mount the servo horn. This flexibility will exaggerate the motion of the servo's movements. If you find a setup too floppy or too wavy, you can use a slightly thicker piece of wire or shorten the length of wire that you are using for a slightly stiffer setup.

At this point, you can take the server with the wire attached and remount it to the servo that is already mounted to the box. As a safety precaution, I suggest putting a piece of painter's tape over the sharp end of the wire to avoid accidentally poking yourself with it until you figure out what you want to mount on it. For this example, I am choosing to mount a flag to the end of the wire.

FINAL ASSEMBLY

Whenever I build a box like this, I find myself fiddling with the length of the wire and the placement of the servo horn relative to the 180-degree arc of the servo. This means I will be yanking the server horn on and off repeatedly as I fine-tune the setup. For this reason, I would suggest keeping the center mounting screw off until you are done fine-tuning the setup. There is usually enough friction on the servo horn to hold onto the shaft without it falling off during testing.

Add a Flag–or Whatever You Want to Wave!

As you can see on page 52, I adjusted to the length of the wire so that there was some flexibility based on the weight of the fabric flag on the end. Aesthetically I found this to be quite beautiful as the servo waved the flag from right to left. The motion was exaggerated by the flexibility of the wire.

Program the Flag Waver

Page 54 shows the program which will allow us to actuate the servo from one position to another position every one-tenth of a second. Earlier in the chapter I challenged you to think about what part of the code from the last program you would change to set an output from pin #12 to pin #3. The blue arrow on page 54 is pointing to a green number that you need to change to do that. Take a few minutes to note how this code (labeled New Code) differs from the previous chapter (labeled Old Code). Compare them line by line. Look for what is similar in structure, then examine the code for differences. Do not get hung up on syntax. Instead, pay attention to structure and see if you can translate into plain English what the new and old code have in common. Do not proceed to the next paragraph without spending some time with this. What you discover on your own will be remembered much longer than me simply telling you the answer!

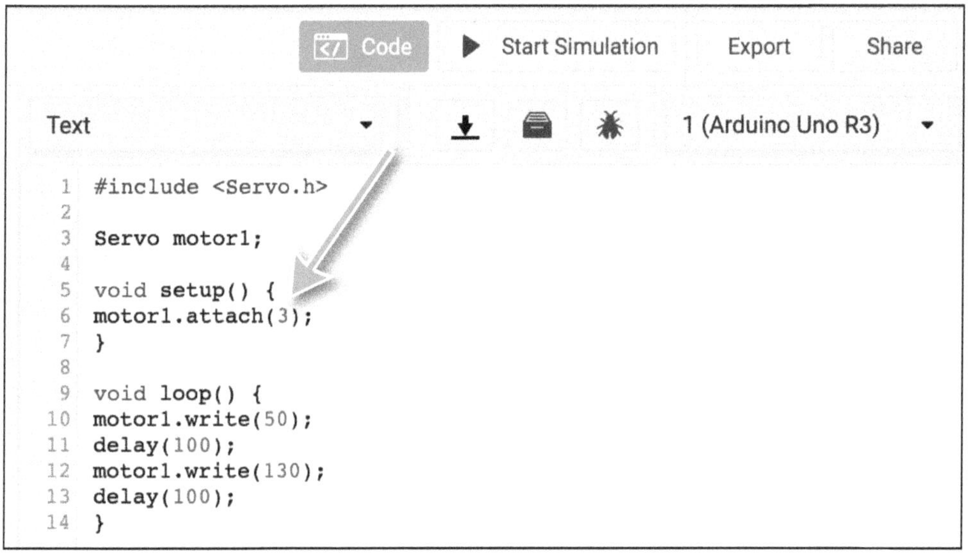

Code Start Simulation Export Share

Text 1 (Arduino Uno R3)

```
1  #include <Servo.h>
2
3  Servo motor1;
4
5  void setup() {
6  motor1.attach(3);
7  }
8
9  void loop() {
10 motor1.write(50);
11 delay(100);
12 motor1.write(130);
13 delay(100);
14 }
```

New Code

Ch3_Light_Switch | Arduino 1.8.9

Ch3_Light_Switch

```
int ledPin = 12;
void setup()
{
  pinMode(ledPin, OUTPUT);
}

void loop()
{
  digitalWrite(ledPin, HIGH);
  delay(1000);
  digitalWrite(ledPin, LOW);
  delay(1000);
}
```

Done compiling.

```
Sketch uses 694 bytes (8%) of progr
Global variables use 9 bytes (1%) o
```

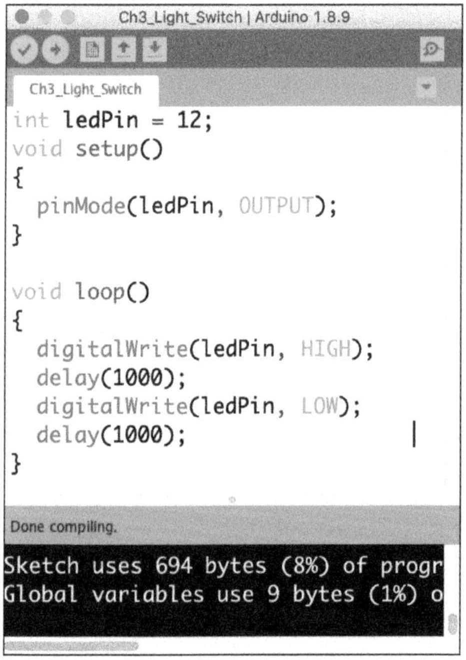

Old Code

Here is what I am hoping you came away with:

1. The first line of code "#include <servo.h>" is unfamiliar to you. This is something called a "library," which is basically a collection of preset commands that can be called in the background of the program. I do not want to delve too deeply into this. For the moment, I will say it is very similar to the way a "digital assistant" enables different skills so you can do things like listen to the radio on your smart speaker.
2. We are setting up the Arduino to understand that there is a servo motor attached to pin #3 rather than an LED attached to pin #12.
3. Instead of a one-second delay (1000 milliseconds) we now have a $1/10$-second delay (100 milliseconds).

If you noted those differences, and understand the meanings behind them, then you are well on your way to writing and modifying code!

WIRING DIAGRAM

The image on page 56 is the wiring diagram for this project. You can see the white wire running from pin #3 to the orange wire (signal) on the servo. The blue wire is running from GND (ground) to the brown wire on the servo, and the red wire is running from the 5-volt to the central red wire on the servo, providing power. This is a relatively standard wiring scheme for servos, although some servos use slightly different colors for the wires. The next-most-common scheme would be for a black wire to represent GND, a red wire to represent 5 volts, and a white wire to represent signal inputs. The general rule is the center wire on the servo plug is usually power, the darker wire on the side is usually ground, and the third wire (typically of a lighter color) is usually the signal. At these voltages there is very little risk of damage to the Arduino or the servo, so feel free to fiddle around with the wires to see what works. If the servo is acting "funny," it is probably because two of these wires are mixed up.

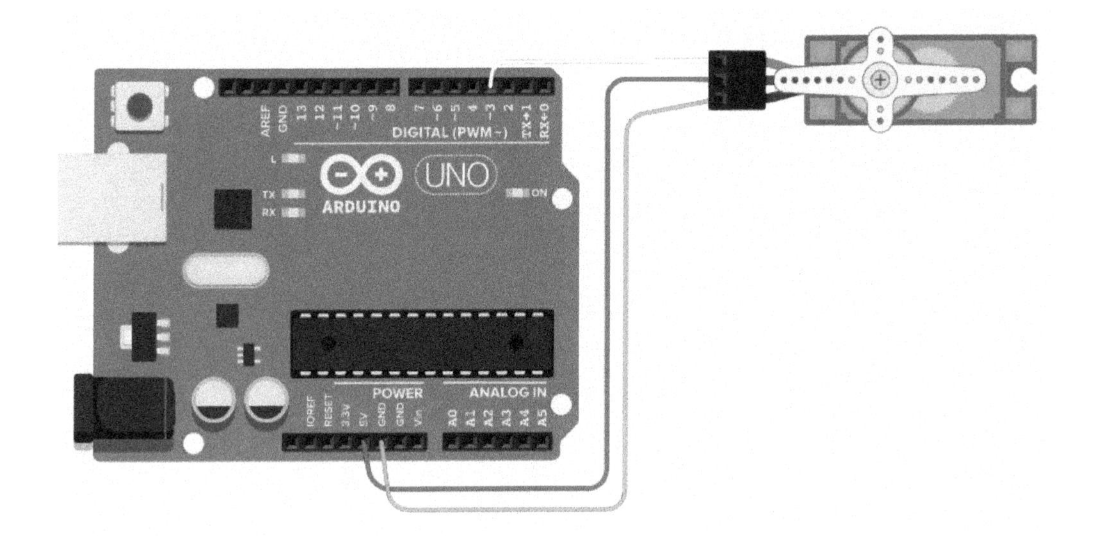

IN THIS CHAPTER YOU:

- added motion to an Arduino project with a servo.
- learned why a servo has three wires.
- wired a servo.
- used a loop in code to command the position of the servo.

TAKING IT FURTHER

Is there a way you can combine the program from chapter 3 and this new program to make an LED blink on pin #12 while the servo moves? Can you change the number of degrees or timing for this current program? Getting to the point that you can improve on existing programs is the best way to get better at coding. While you are doing this, please explore the example programs in the Arduino IDE. If you really want to stretch your learning, explore the iteration below!

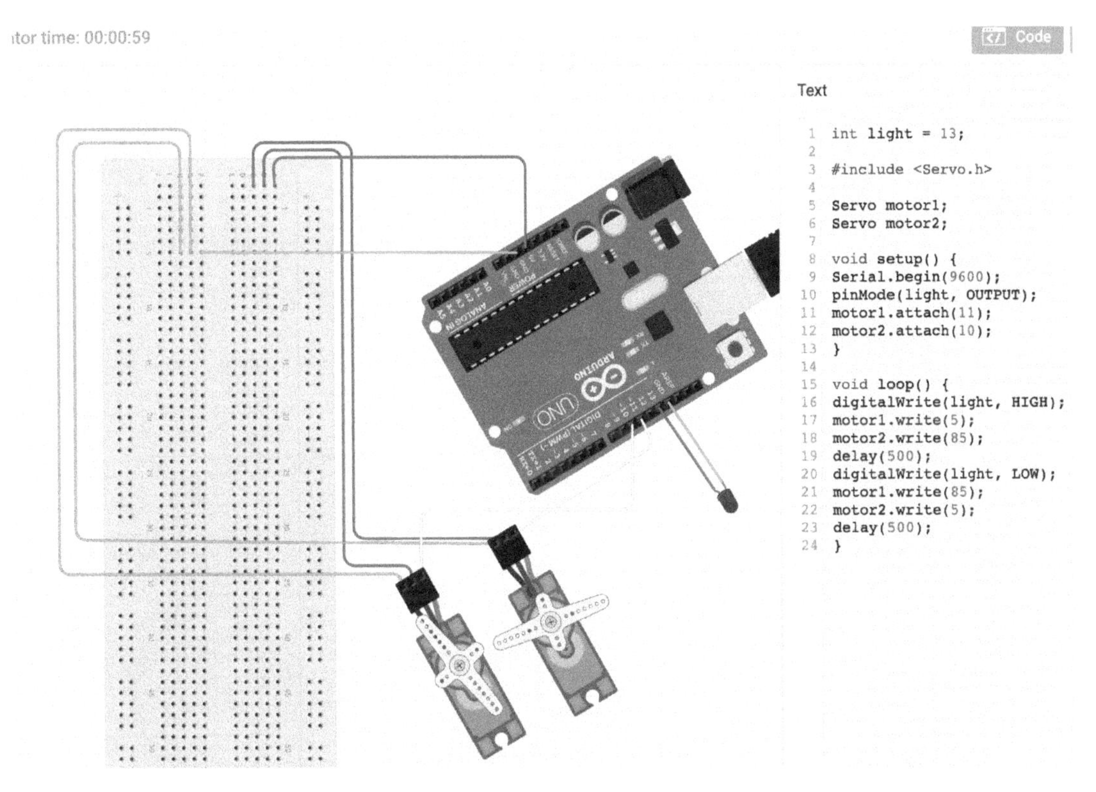

Do an Internet search for "PCA9685 16 Channel PWM" for servo drivers that can control much larger servos (and many of them). Larger servos need more power than the Arduino has to provide but the wiring is fairly simple after you read the disco shoe project. Larger "standard" servos are typically rated as "13kg." Always get metal-gear servos when possible.

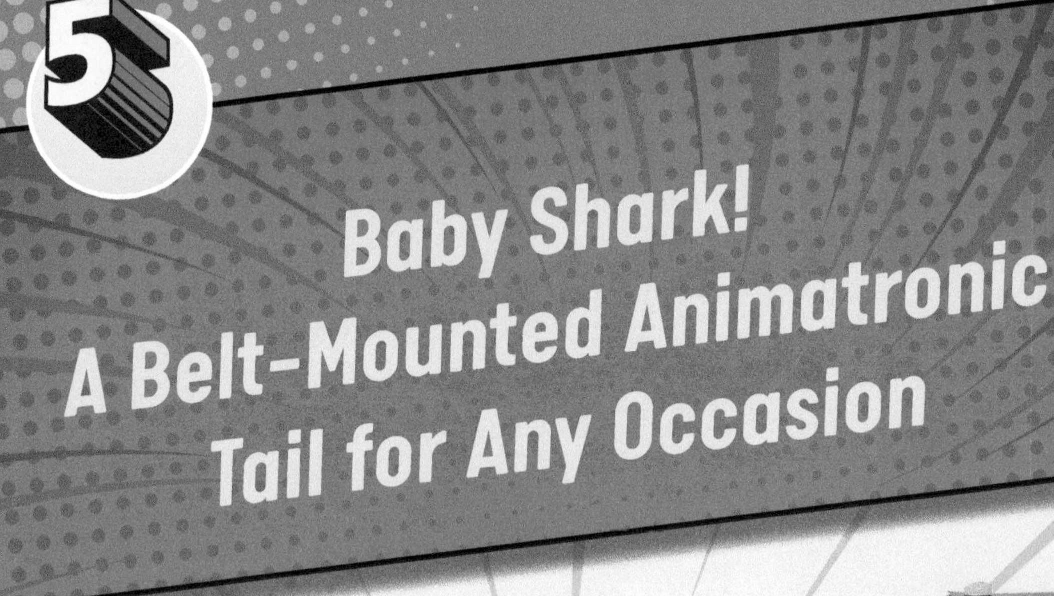

5 Baby Shark! A Belt-Mounted Animatronic Tail for Any Occasion

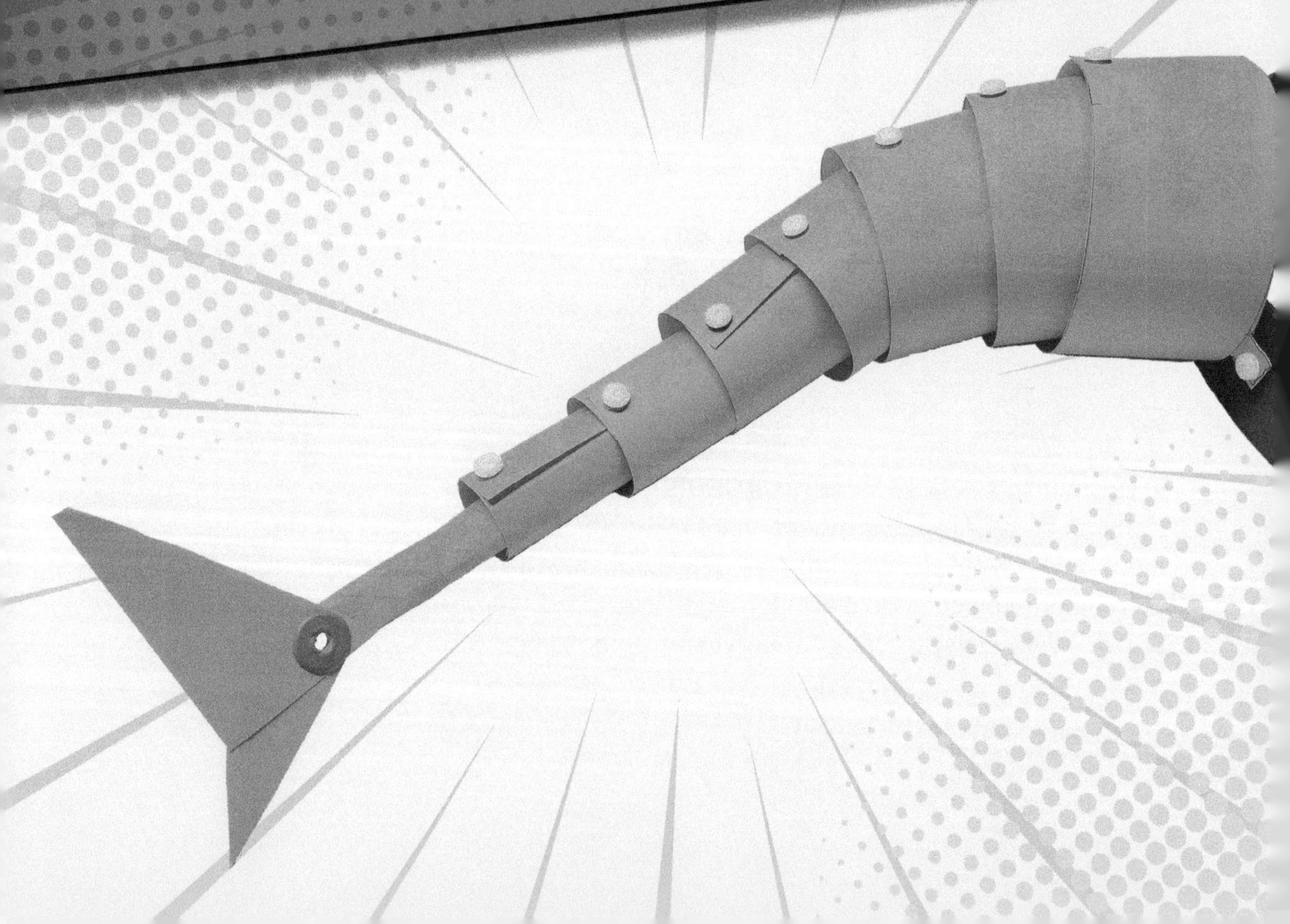

MATERIALS USED IN CHAPTER 5:

- Used Leather Belt, 1.5- to 2-inch wide
- Crop-A-Dile Eyelet Tool
- $^{5}/_{16}$-inch Eyelets with Wide Head (20)
- 0.031-inch Piano Wire
- 4mm x 1mm Nylon Screws (20)
- 4mm Nylon Nuts (20)
- Male-to-Male Breadboard Jumper Wires (3)
- 6-inch Zip Ties (3)
- 12" x 12" Textured Cover-stock Paper
- Scissors
- Arduino Uno
- Hard Wire Cutter
- Small Needle-Nose Pliers

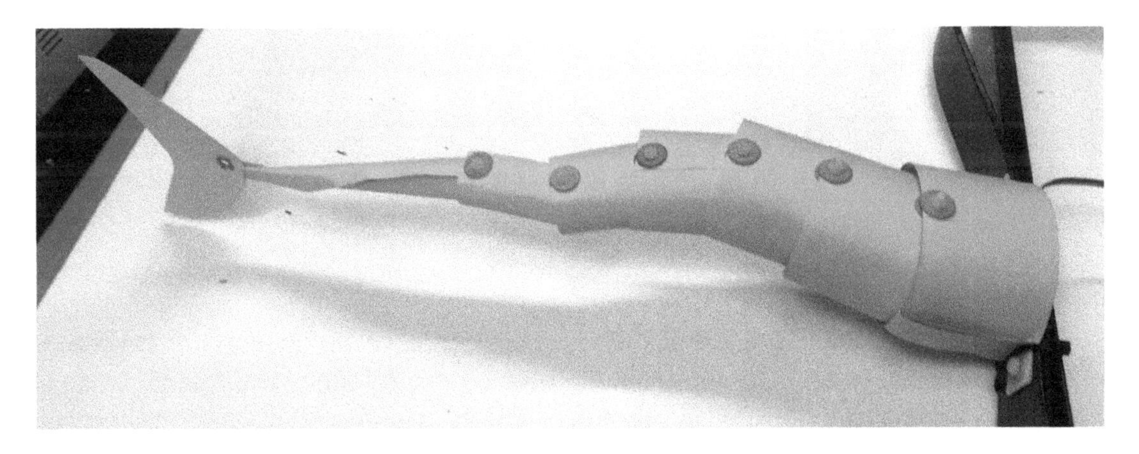

Never, ever walk into a second grade classroom and start humming "Baby Shark" by Pink Fong unless you know what you are in for. As of 2019, the song has racked up more than 2.8 billion views on YouTube, and it's a teacher's worst nightmare. It is a contagious song you can never un-hear, even as an adult. If you sing a line of the song to a classroom of second graders, you will set off events that cannot be contained! In this project you will build your very own wearable animatronic "baby shark" tail. Having built and actually worn it, I can say it is both fun and a little disturbing. To complete this project, we will need a few new tools and materials.

Total project cost: $15.

From prior projects in this book you already have the Crop-A-Dile eyelet tool, the Arduino Uno, and the jeweler's pliers with conical jaws. In this chapter you will learn about piano wire and how to use it as a push-pull element of your animatronics. Piano wire is similar to an unfolded paperclip, but the steel it is made from is a steel that is flexible and will always return to the same shape.

For this project, we require a new tool called a hard wire cutter. Piano wire can be cut by a standard wire cutter; however, the user runs the risk of dulling the wire cutter very quickly and possibly damaging the jaws of the cutter. A hard wire cutter uses a shearing motion to cut the piano wire and does not get dull with use.

A Hardwire Cutter

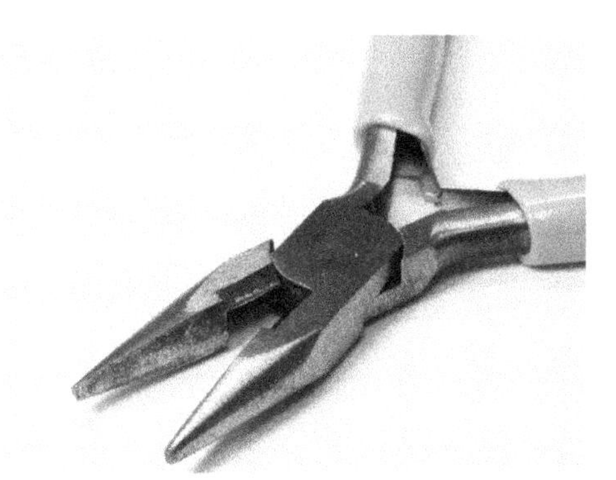

Small Needle-Nose Pliers

Using needle-nose pliers, bend a very small loop in two pieces of piano wire to create a pivot. Bending piano wire can be a bit more finicky than paper clips. Using the ⅛-inch diameter wire, the first section of needs to be cut to 18 inches, with the loop in the middle of the wire. The section of wire is approximately 9 inches long and needs a similar loop at one end. The key is to use your finger to push the wire as close as possible to the very tip of the needle-nose pliers without losing your grip on the wire as it is bent around the tip.

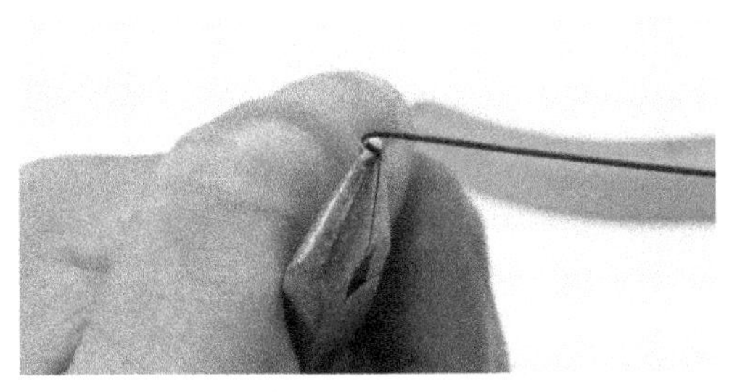

Start the loop

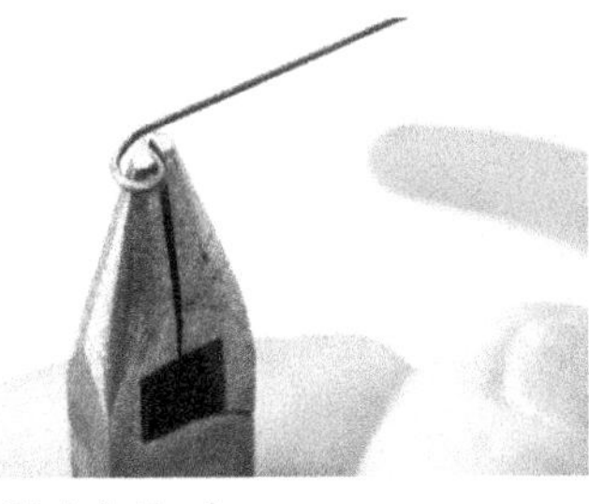

Finish the loop

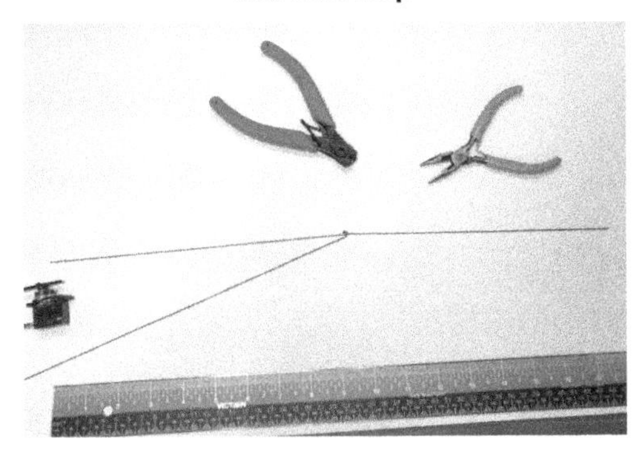

The connected 9" and 18" wires

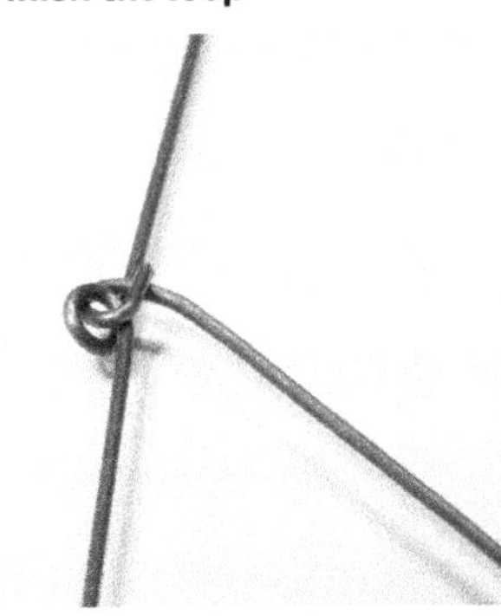

The two wires joined at the loops

ATTACHING THE BELT TO THE SERVO

Having a wide belt is essential. The stiffness of a wide belt will ensure that the shark tail is held rigidly out from the body and does not sag.

Steps:

1. Lay the belt on a flat surface, with the side intended to face the body facing up.
2. Use a permanent marker to outline the location for the servo.

3. Use the 3⁄16-inch punch on the Crop-A-Dile to make holes.
4. Run the zip tie behind the belt from the front.

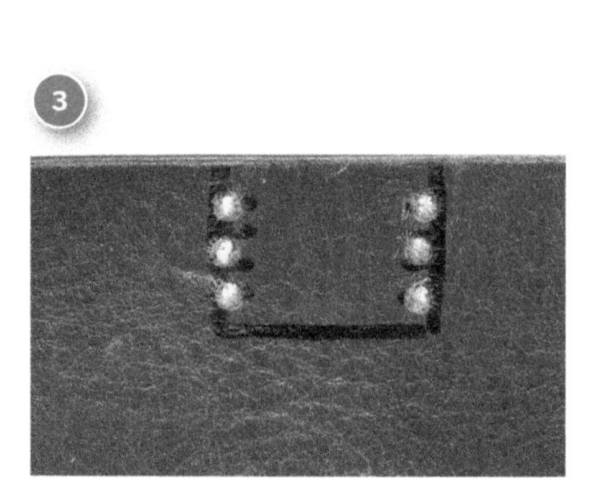

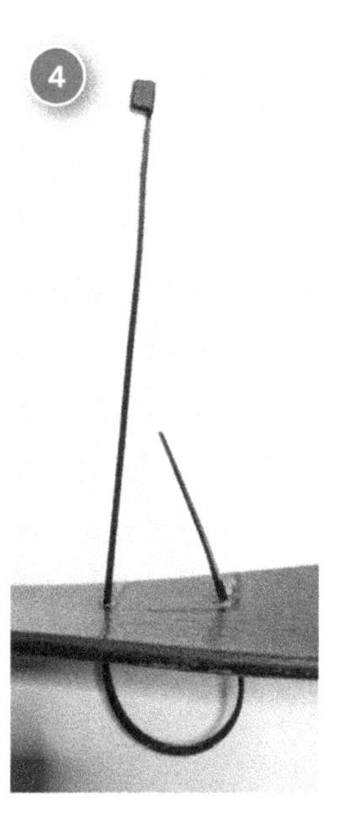

5. Add two more zip ties.
6. Make sure the wire goes between two of the zip ties.

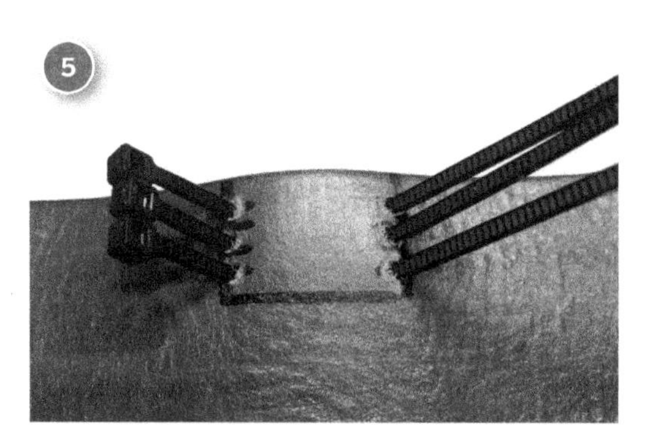

7. Check your build against the photo below.

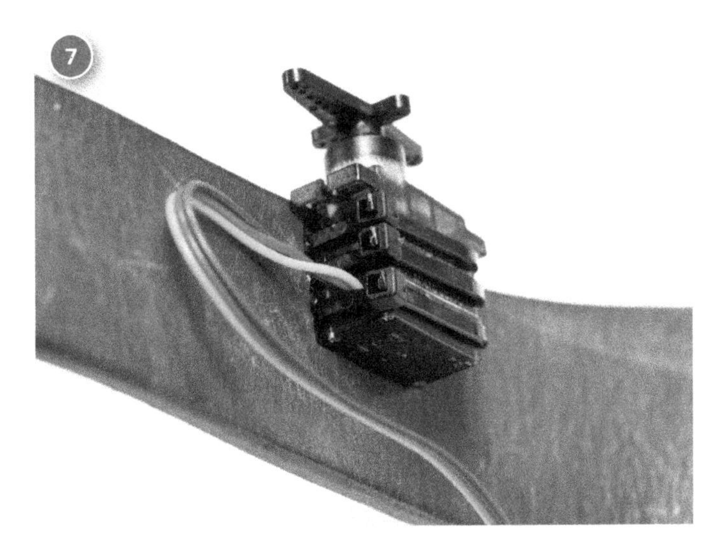

MAKE THE TAIL SKIN

This project uses a small standard "micro" 9-gram servo. These servos are typically available in several varieties. Some have metal gears, and others have plastic gears. Both come in 360-degree and 180-degree varieties. Choose a 180-degree metal servo for this project. Because we are using a smaller servo, we will need to keep weight in mind as we build this project. That means we'll design a smaller "flicky" tail rather than a longer, thicker, and heaver tail. The best material I have found for this project is textured cover stock. Glitter stock is a good choice too, but it is harder to roll, so I would suggest starting with the slightly lighter textured cover stock. It comes in 12″ x 12″ pads in different color palettes. Students in my class demanded pink for the preferred color of baby shark, so we went with that. We are building a tail that is between 16 and 18 inches long using a single sheet of paper. Use a ruler and permanent marker to lay out the pattern on page 64 and then cut those pieces out with scissors. This will leave a 2-inch x 12-inch piece of scrap left over that we will use to thicken the tabs in the next steps.

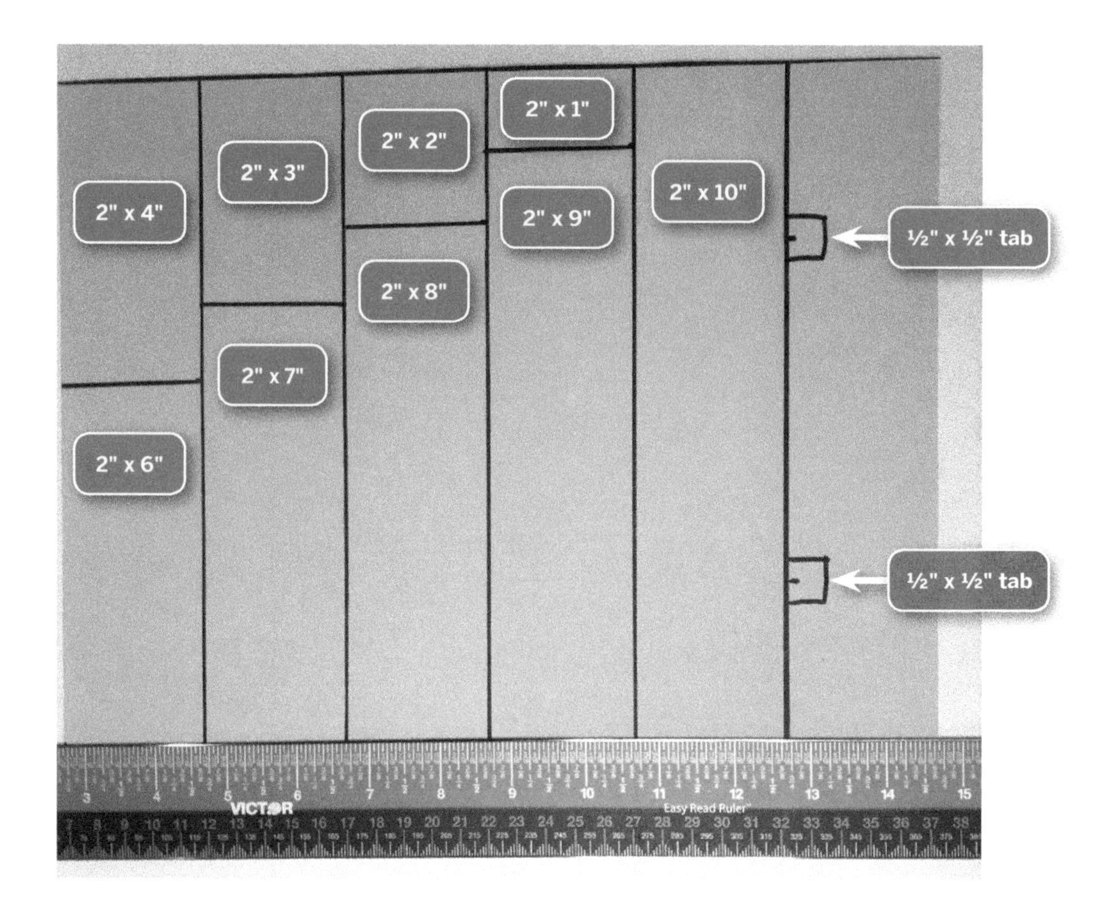

THE MISSING LINK

As an engineer and mechatronics teacher, I would argue that there are several differences between folks who feel comfortable making animatronic projects and those who struggle to make machines that work as well as they imagine. The point of this book is to walk the reader through a series of tools and techniques. The reason I am reiterating that point in this chapter is that a very important technique is used in the steps below. I see too many people taping, gluing, and using other shortcut techniques to attach parts to a servo. There is a correct way to do this, and that is with 0.031-inch piano wire, and more specifically, creating a sharp S-bend at the end of the wire. I suggest that you make several test bends until you can get results as close as possible to two opposite

90-degree bends in the wire. The vertical part of the wire in the photo needs to be exactly at 90 degrees to the servo horn (the cross-shaped bit of plastic that comes with the servo).

Steps:

1. Make the S-bend precisely with sharp angles.
2. Twist and insert the wire into the servo.

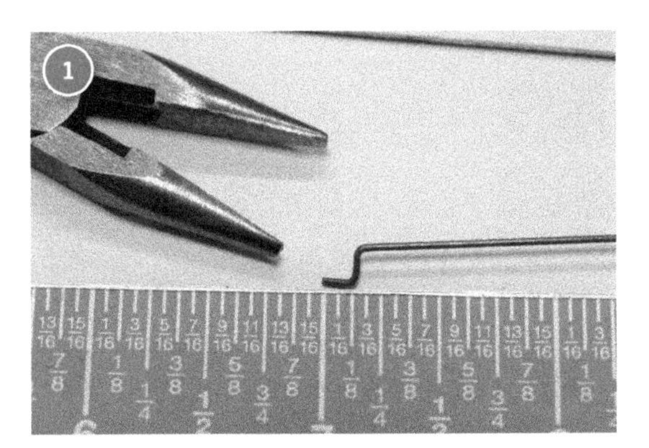

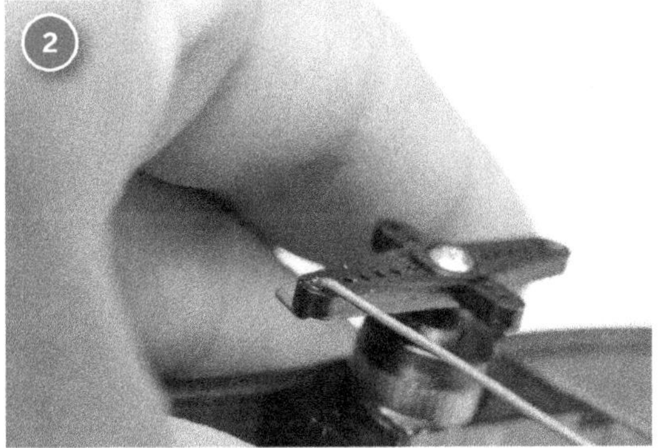

3. Rotate the wire down so it snaps into the hole (in this case the wire would have been pulled down toward the bottom of the page).
4. Mark the other end of the other wire where it crosses over the servo mounting hole.

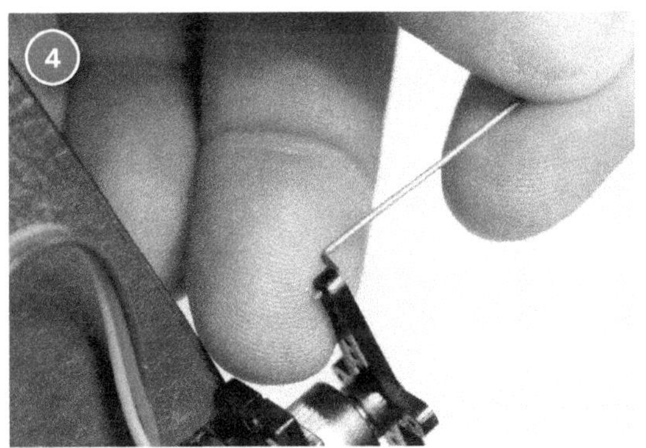

5. Bend the wire around that first plastic tab with the mounting hole.
6. Wrap it a couple more times. This can be less precise than the S-bend.

CROP-A-DILIN'

We started using the Crop-A-Dile in chapter 2 to connect paper straws to craft sticks. In that case we used one eyelet to connect both objects. In this case we will use one eyelet on each part, so we have two metal holes that line up for a strong, almost friction-free pivot. We will use nylon screws to connect each joint. For now, note that every paper section has four eyelets (two in the front and two in the back, along the center lines of the baby shark tail). Spacing and precision are not super important here. I space the eyelets so that they barely touched the edge of the paper. These large eyelets can get stuck in the tool if you accidentally invert the jaws of the tool. Remember that the steel tooth goes into the top (fat) part of the grommet.

Tooth goes to top of eyelet

Mound goes to bottom of eyelet

Steps:

1. Roll each section of the tail skin over, allowing for approximately 7/16-inch overlap. If you want to get fancy, then roll each section into a subtle cone shape with about 1 degree of taper. However, if that is too complex, don't worry. The tail looks great even without the taper.
2. Use the 5/16-inch hole punch on the Crop-A-Dile to punch a hole in either end of each paper tube.
3. Set each eyelet using the Crop-A-Dile. Do this to all sections, from the largest to the smallest section.
4. Do not connect any of the tail sections at this point. You may line them up and check that the holes overlap between sections.

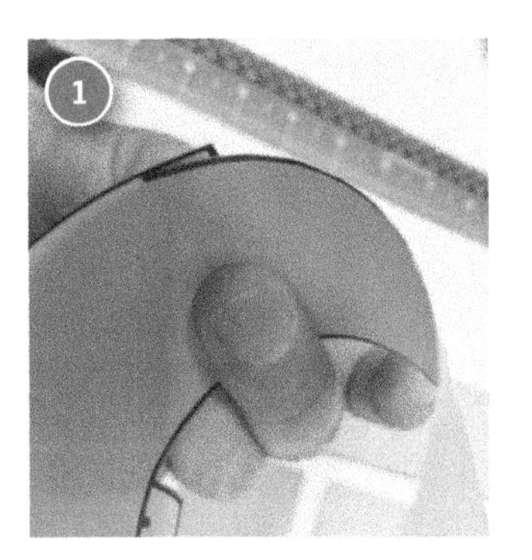

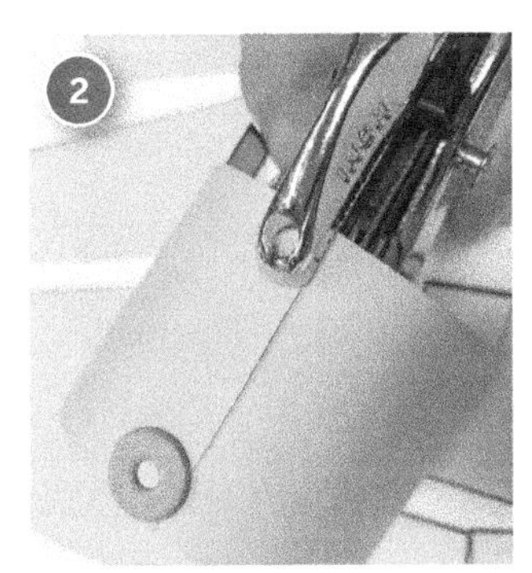

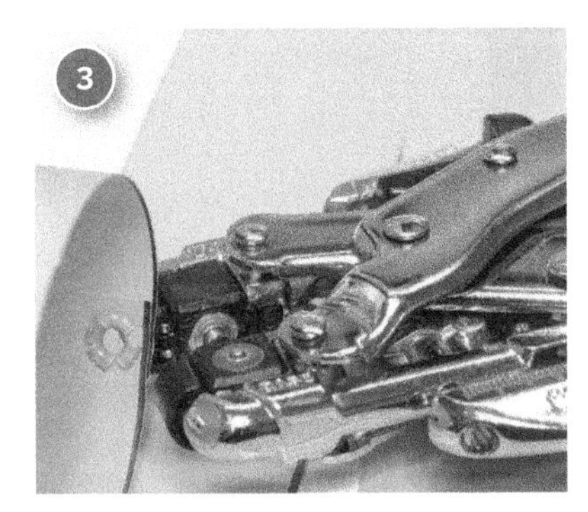

THE BASE OF THE TAIL

Prepare the base of the tail in the same way as each of the tail sections on the prior page. We will now look at how we can best attach the tail to the belt. We will use the "flange" technique. A flange is any feature of a part that connects perpendicularly to another part, usually a cylindrical connection to a flat-plane surface. This kind of joint is common when building structures out of pipe, such as stair rails or playground sets. The two tabs on the 2-inch x 10-inch strip of paper will act as our flange to connect to the belt. The entire weight of the tail will rest on these two connections. Due to the side-to-side motion of the tail, we really need to think carefully about how to reinforce them. In this case, I have chosen to cut ½-inch strips from the section of paper we did not use. We will need two strips to wrap around the full width of the 2-inch tail section, plus the ½-inch square tabs.

Steps:

1. Wrap the ½-inch strips from the scrap paper around both the tab and the full width of the section.
2. Continue to wrap the ½-inch strip around until you have at least three layers over the tab.
3. Punch 5⁄16-inch holes in the center of each tab and through the other overlapping layers of the strip.

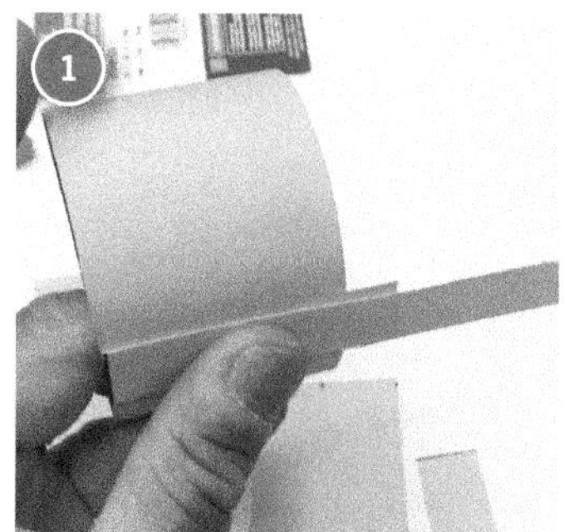

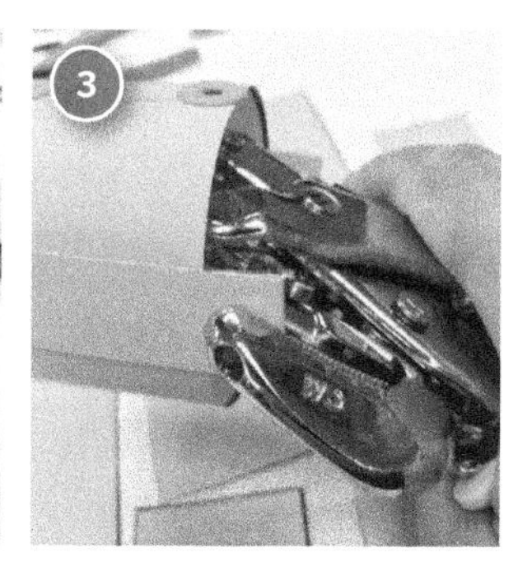

4. Attach the large eyelets to each strip-wrapped tab.
5. Align the base section with the tabs to the belt and the second section to the opposite end.

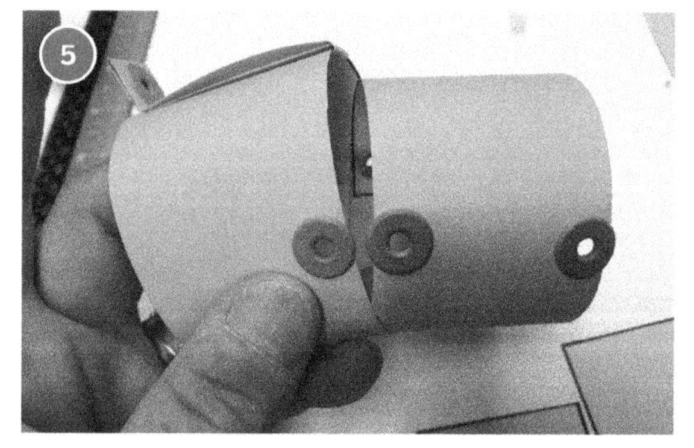

Nylon Screws Option (Build Variation)

This project can be built with two different types of nylon screws. The first version is to use 10mm long M4 nylon screws and M4 nylon nuts. The second version of the build is a little more expensive but adds some build simplicity. It uses nylon rod of the same M4 diameter and pitch (thread spacing). The rods can be ordered in 3-foot sections so you can cut lengths long enough to run though the entire thickness of the tail. This adds around $9 to the build, but you will have plenty of threaded rod left over for other projects.

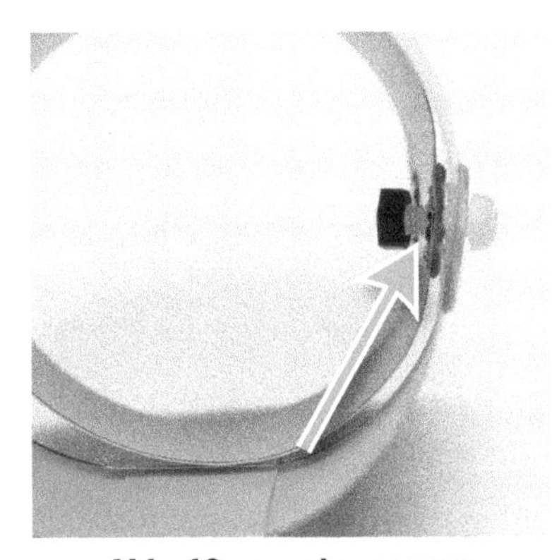

M4 x 10mm nylon screws

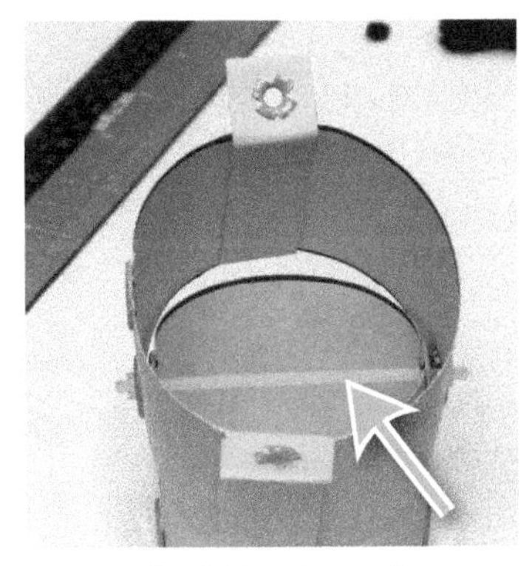

Cut M4 nylon rod

Smash screw ends to lock them in place using the tip of the jeweler's pliers with conical jaws. This can be done on the interior with normal 10mm screws or on the exterior if you are using nylon rods. One of the wonderful properties of nylon screws is that they can be cut without messing up the threads so you can trim and reuse if (when) you need to take apart your project to iterate.

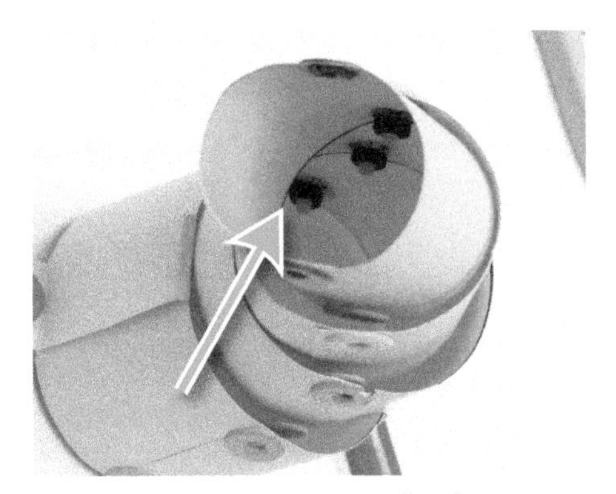

10mm M4 nylon screws installed in top row

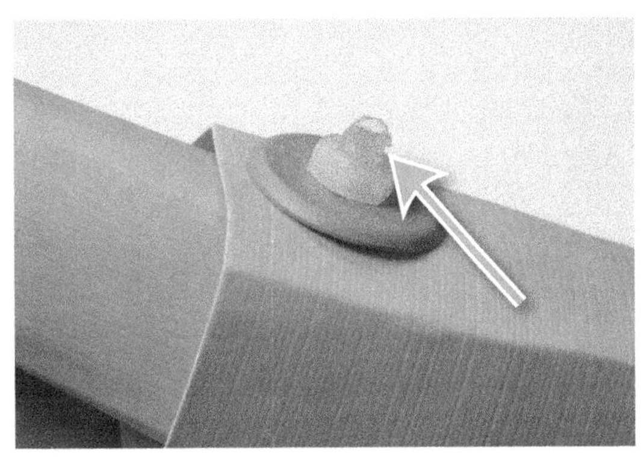

Smashed ends of nylon rod after nut installed

TAIL ASSEMBLY

Steps:

1. Start to assemble the tail using the preferred M4 screw or rods, starting with the larger sections and adding progressively smaller sections, until we get to the smallest section that can fit over the Crop-A-Dile tool.
2. Take a section of scrap paper and cut a trapezoid 2.5-inch x ¾-inch x 8-inch.
3. Roll the trapezoid diagonally so one end almost comes to a point, while the other end (the fatter end to be attached to the tail) is just smaller than the diameter of the next larger tail section. This will make it about ⅝ inch on the wide end.

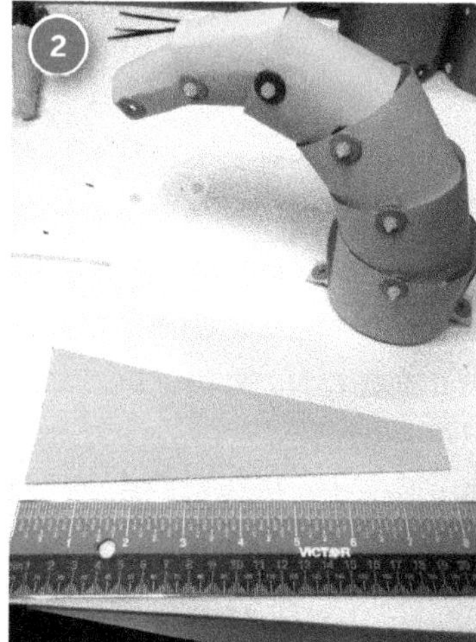

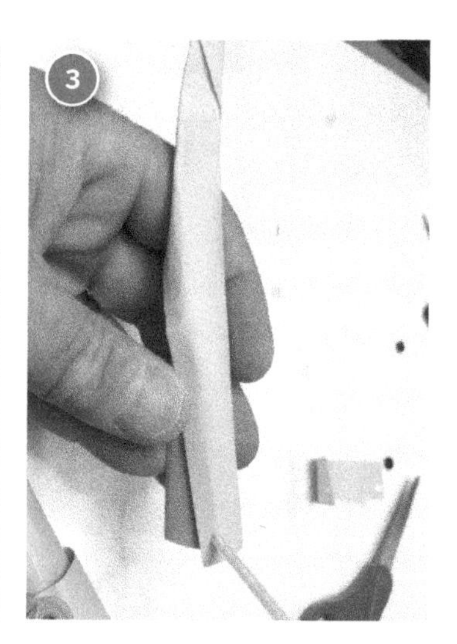

4. To make the pivot, choose one of these options:
   a. Use a single 10mm screw and squash the opening a bit.
   b. Use a section of nylon rod.
   c. Use a 1-inch brass brad.

The assembled tail should be floppy and loose when moving side to side but somewhat rigid moving up and down. This is an exciting point in the build! The tail should have approximately 180 degrees of total travel. To avoid stressing the joints on the paper tail, the skin should have a longer range of travel than the piano-wire section attached to the servo. The total length of the tail at this point will be about 16 to 18 inches.

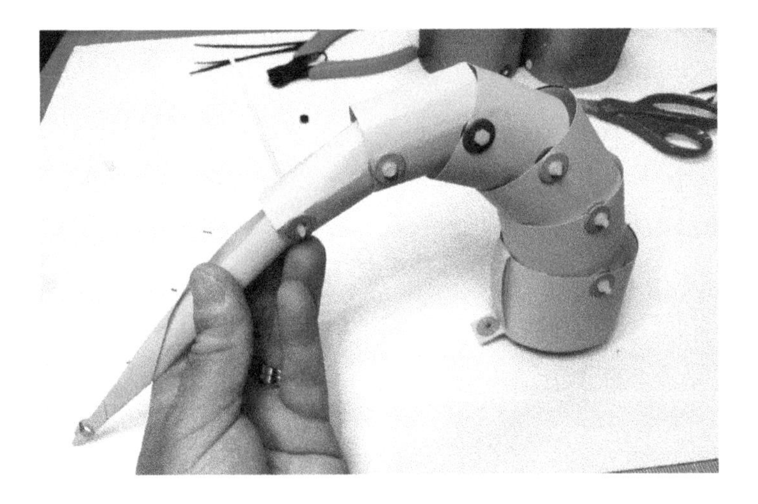

MAKING THE TAIL CONNECT

We will now assemble the tail, skin, and belt containing the servo together. We are just about 20 minutes away from the first tail flick! The first step is to line up the tail skin next to the piano-wire rod. There should be at least a ½ inch or more of piano wire extending beyond the tail. We will trim that in the last step when we add the tail fin.

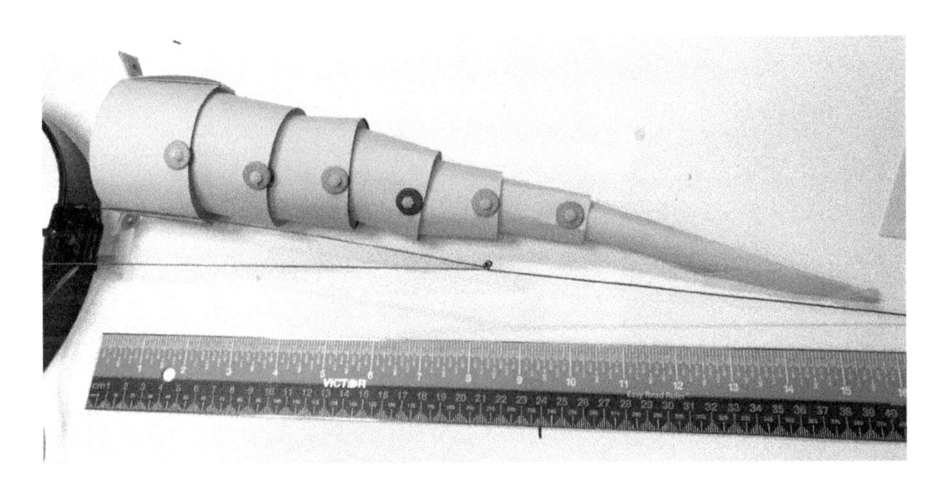

The picture below is a view showing how you would connect the various tail sections if you choose to use nylon rods instead of 10mm nylon screws. In either case, you will place the tail over the servo with the tabs/flanges on the belt on either side of the servo.

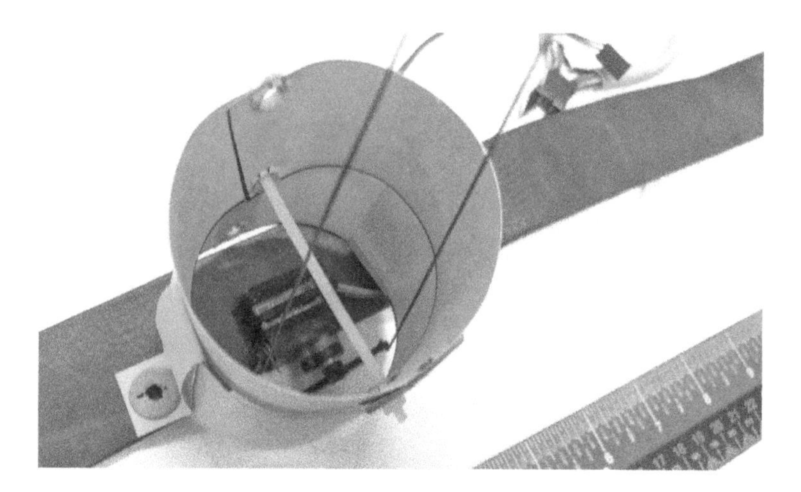

Once you have the skin of the tail over the servo, skip ahead to Tail Sketch on page 75 to make sure you have the servo horn in the correct position. Remember from chapter 1 that you want the servo horn to be in the neutral position when you send the 90-degree command since the servo has a 0- to 180-degree travel. In this case that means that when the servo is approximately parallel on the long section of the servo horn at the 90-degree command, we want to push the code to the Arduino and see exactly how many degrees we want to actuate the servo.

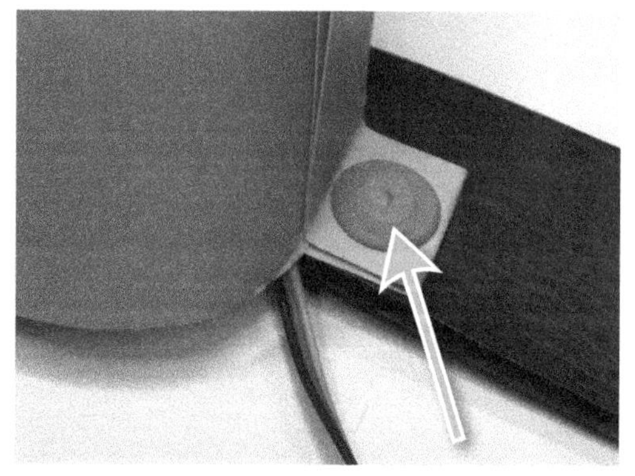

M4 x 10mm screws pass through eyelet and belt

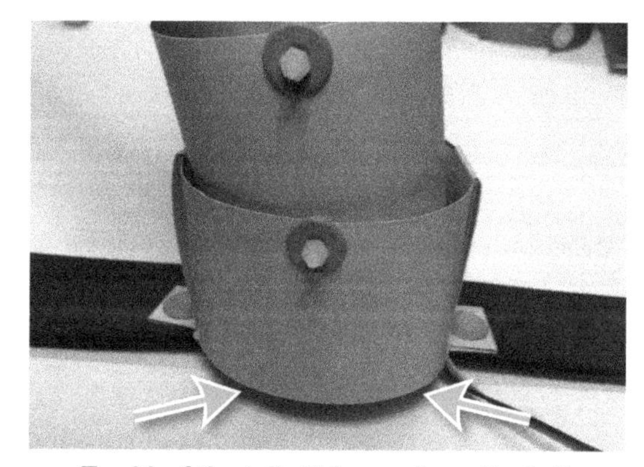

Top 1/3 of the tail sticks up above the belt

The next step is to punch 5⁄16-inch holes in the belt to align with the tab holes and use M4 x 10mm screws. Make sure the servo wire is free to come out of the top side of the belt under the paper.

View from inside of belt

Accidental preview of GlitterShark Tail

Hasty Hole (oops!)

M4 nylon nuts backing the screws

THE CODE

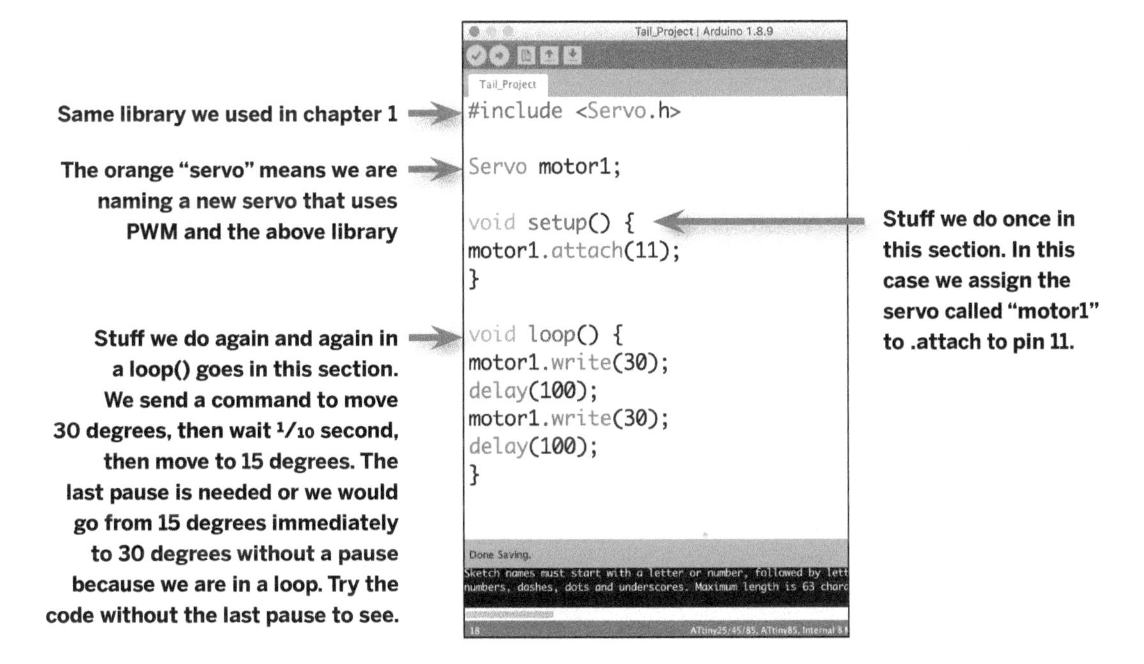

I test both the wiring and code of all my Arduino projects using the circuit simulator inside of Tinkercad. This is the section in Tinkercad that allows you to both virtually wire a circuit as well as run the code to see how the circuit will behave. Note that we are still using the 5v and ground from the right side of the Arduino and getting a signal out of the left side on pin #11. The pins with the wavy lines next to them are the pins that can send a PWN (pulse width modulation) that the servo needs to function. We will cover PWN in the next chapter. For now, wire the Arduino as shown.

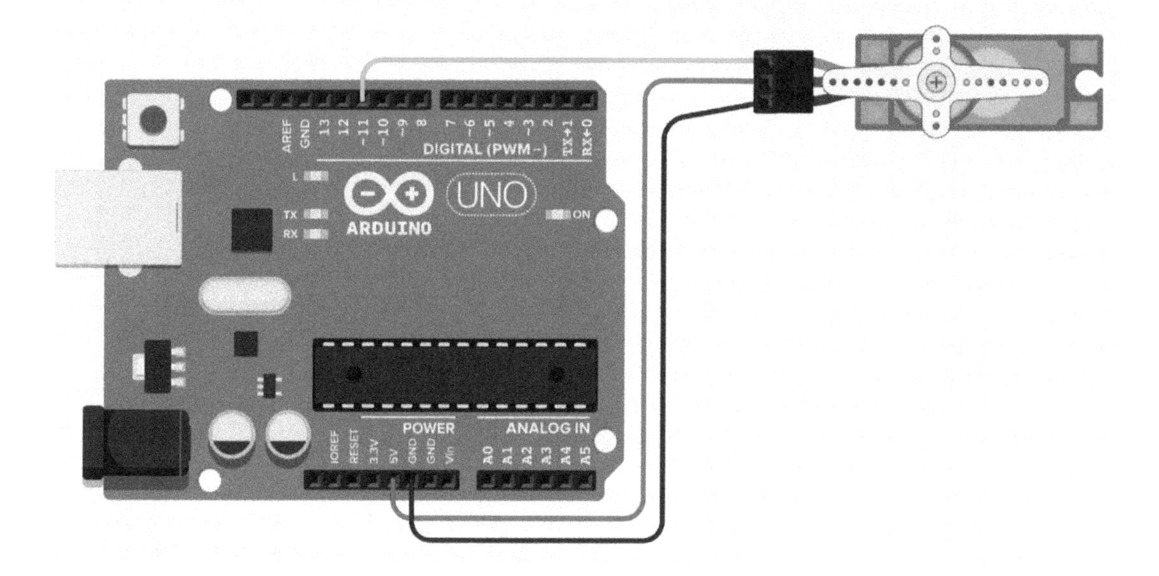

The wiring uses male-to-male jumper wires to connect the servo as follows:

1. Ground to Ground
2. 5V to the servo power (center wire)
3. PWN Signal from pin #11 to the 3rd wire.

Most servos, even cheap ones, are pretty robust. Do not worry if you mix up the wires. If your code is loading to the Arduino and the servo is not moving, then check the wiring.

ADD A TAIL FIN!

The tail fin can be cut from a single piece of scrap paper. Use the template below to trace a tail or make one of your own. The tail is proportional when it is about 3 inches tall. This is where we left some of the piano wire intentionally sticking out so that we can use an eyelet to attach the tail. The piano wire should remain loose to enable the tail to extend and retract as it flicks from side to side. The piano wire can be trimmed to the end of the tail fin. (Be sure not to have any sharp portions protruding that could injure someone.)

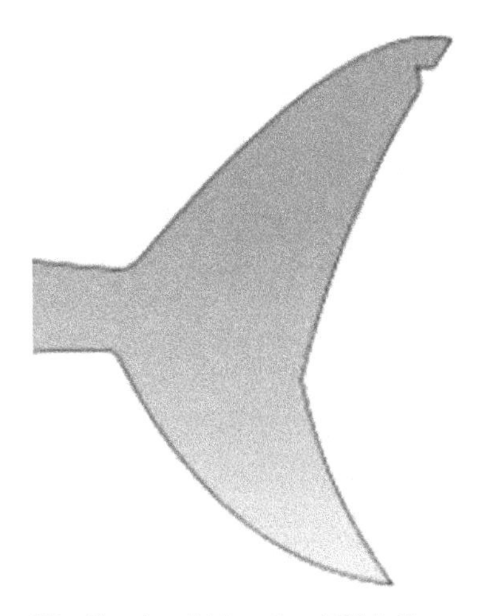

The fin should be about 3" tall

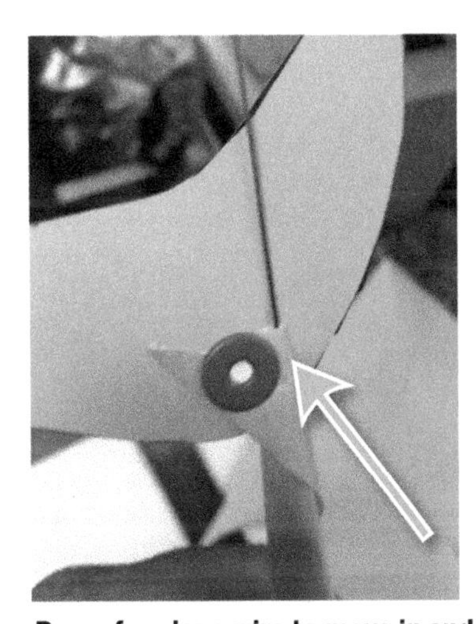

Room for piano wire to move in and out

FINISHED!

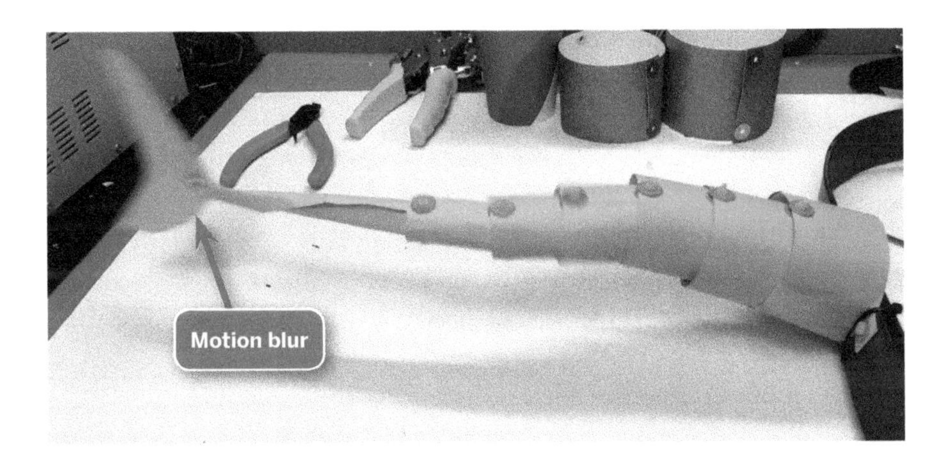

At this point in the book you should be calling yourself an "Arduinist" instead of an Arduino user. (In other words, you are coming along quickly.) You are now at a point that took me three years of stumbling along to reach. You are literally building a whole machine: chassis, integrating electronics, and coding. The difference between you and a company full of engineers building a generation of self-driving cars is mere levels of detail. That is not an overstatement. You are now engineering.

IN THIS CHAPTER YOU:

- built a chassis using the Crop-A-Dile tool.
- built the circuit that connects the Arduino to the servo.
- uploaded code to bring the whole machine together.

TAKING IT FURTHER

Do an Internet search for "heavy eyes shark" or go to HeavyEyes.Co and look up the shark kit. Matt Cavanaugh put together an amazing toy shark kit that spawned a lot of what I did in my classroom last year. In putting the kit together, I leaned to really understand the beauty in 2-D geometric net to 3-D design. To make larger cardboard versions of the shark, the students simply took pictures of the parts laid out and then converted them to vector files. If you are interested in Cricut or Silhouette cutters or are looking at a laser cutter like the Epilog or Glowforge, then this is a really fun kit to digitize and riff on. Keep in mind that Matt put a lot of work into the design, so be kind and buy a kit from him if you are going to digitize and riff on the design.

6
Disco Shoes

MATERIALS USED IN CHAPTER 6:

- **Adafruit NeoPixel Ring (12 pixels max)**
- **26-gauge Silicone Wire**
- **0.039-inch or 0.032-inch Music Wire (or a Heavy-Duty Staple)**
- **Hard Wire cutter**
- **Small Needle-Nose Pliers**
- **Jeweler's Pliers**
- **6 x AA Battery Holder with 2.1mm x 5.5mm Barrel Jack**

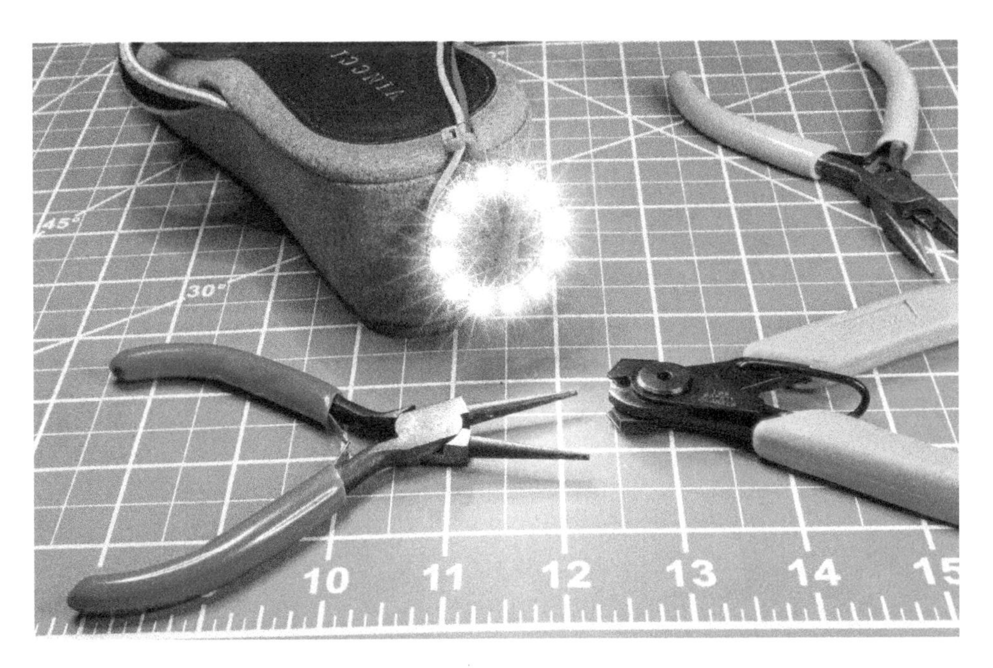

Whenever I can build a project that makes somebody else smile, I get a lot of satisfaction from my efforts. For this project, I took an old pair of shoes my wife had put into the donation pile and modified them to be true standouts on the disco floor! The lights are blazingly bright and shift through all the colors of the rainbow in any pattern I want to program them. The reason I am including this project in the book is to build on the knowledge you gained in chapter 4 as the newest member of the three-wire club. In that chapter you learned that the third wire carries a signal to whatever device we attach to the Arduino in addition to the positive and negative wires used for power. In the Taking it Further section of that same chapter, I included a diagram for running more than one servo at a time. In that diagram, I used the breadboard to connect multiple power and ground wires together because the Arduino does not have enough 5v power and ground pins to hook everything together. With more complex machines, you can quickly imagine the maze of wires that this would create.

What if we could send signals meant for different devices along the same signal wire at the same time? In this case we will use a ring of 12 LEDs (do a search for "Adafruit NeoPixel Ring - 12 x 5050") and just three wires to power and control it all!

The rest of the projects in this book simply iterate on more complex motors, inputs/outputs, and using smaller wireless variations of the Arduino. When you finish this chapter, you will be ready to go exploring!

Use the steps below to wire the NeoPixel ring.

1. You will use three of the four connection points on the ring. PWR is where you connect the 5v from Arduino. GRD is going to GND on Arduino, and IN is where you attach the signal wire.
2. Cut the end off of jumper wires or use a roll of wire cut to length. (My favorite wire that is super easy to strip with your fingernail and easy to tie is pictured in image 6 on page 81.) In all cases, you will strip off about 1.5 inch of wire insulation from one end and then twist the small exposed metal fibers into a very tiny rope to prevent them from fraying.

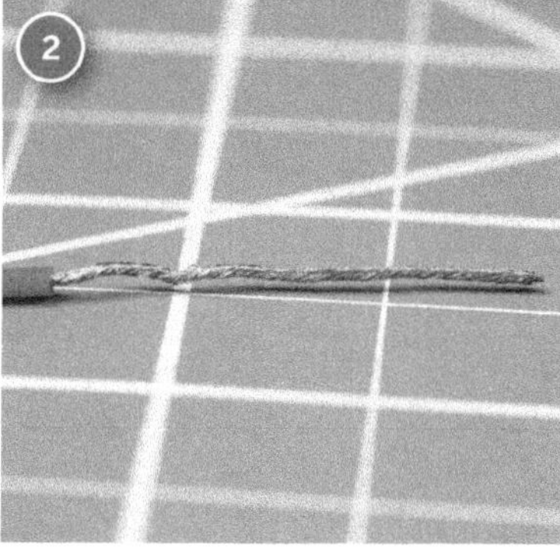

3. Thread each of the three wires through their corresponding holes one at a time.
4. Knot each one by wrapping the extra exposed wire a few times around the part coming in. Cut off the extra wire after a few wraps. One key here is adding tension as you wrap the wire back around itself. Another key in #3 is to leave a bit of distance between the insulation and the circuit board so when you wrap the wire back around on itself, there is metal-to-metal contact on both the wire and the small copper ring that makes up the hole in the circuit board. Your goal is to make an electrical connection between the two.

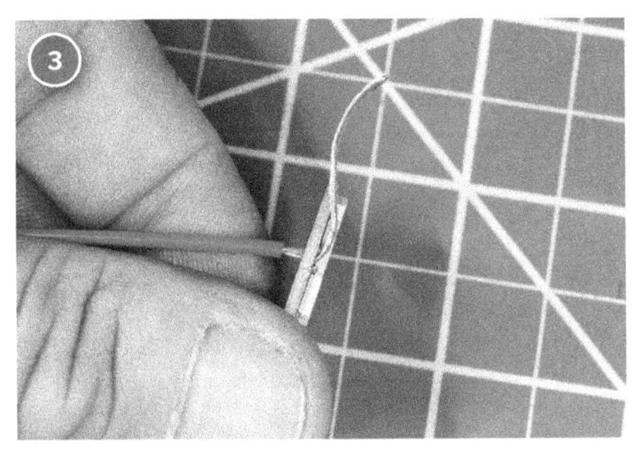

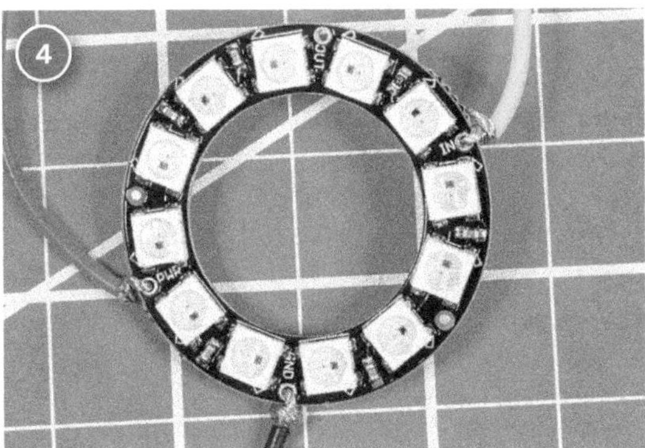

5. You can see all three wires connected; in this case, I used yellow for the signal wire. The OUT is a place you can extend the string of lights.

Wiring Note

If you choose to cut off the end of a jumper wire and strip and twist it on the last step, then you do not need the steps below. If you choose to use silicone wire, then you will need to attach the other end to a jumper wire, then you must:

1. Use jeweler's pliers to make a small half round hook at the end of a jumper wire.
2. Wrap the silicone wire around the hook and cut off the excess.
3. Connect the jumper wire to the Arduino.

FUTURE PRECAUTIONS ON BIGGER PROJECTS

There are some other components called resistors and capacitors that you should include when working with larger numbers of NeoPixels. The purpose of the project in this chapter is to get you up and running quickly versus bogging you down with a ton of detail about voltage spikes, electrical noise, and resistance changes. My job here is to present the quick-and-dirty approach to get you started. (Check out www.adafruit.com and www.instructables.com because both provide more in-depth explanations.)

Larger Power Supplies

Earlier, I mentioned that a 9-volt battery was the worst possible choice for an Arduino power supply because, although it has the correct amount of voltage, it does not have the capacity to provide the amperage/current (number of electrons) to the Arduino that is necessary to power a number of lights or to provide the power required by a large motor. The common element on all of these power supplies is the 5.5mm 5.5-barrel connector on the end of each. The Arduino has a female connector, and our goal is to put 7 to 12 volts DC from a battery or battery charger into that. A search on Amazon for "5.5 barrel" combined with "Arduino" or "AA" "power" will bring up a lot of options. A few are shown on page 83:

- The worst choice, but the easiest to set up (#1).
- A covered battery box with an integrated switch that holds 6 AA batteries (#2).

- A less expensive battery box that also holds 6 AA batteries and does not have a switch (**#3**).
- An in-line switch that you can add between a battery and the Arduino (**#4**).
- This "barrel jack" has the green part that is exactly the same mechanism on the side of the relay in chapter 3 (**#5**).

The idea here is to cut the end of the wires off a 9-volt or 12-volt battery charger that plugs into the wall and use that as a power source. When you do that, be sure you know how to use a voltage meter, which is beyond the scope of this book.

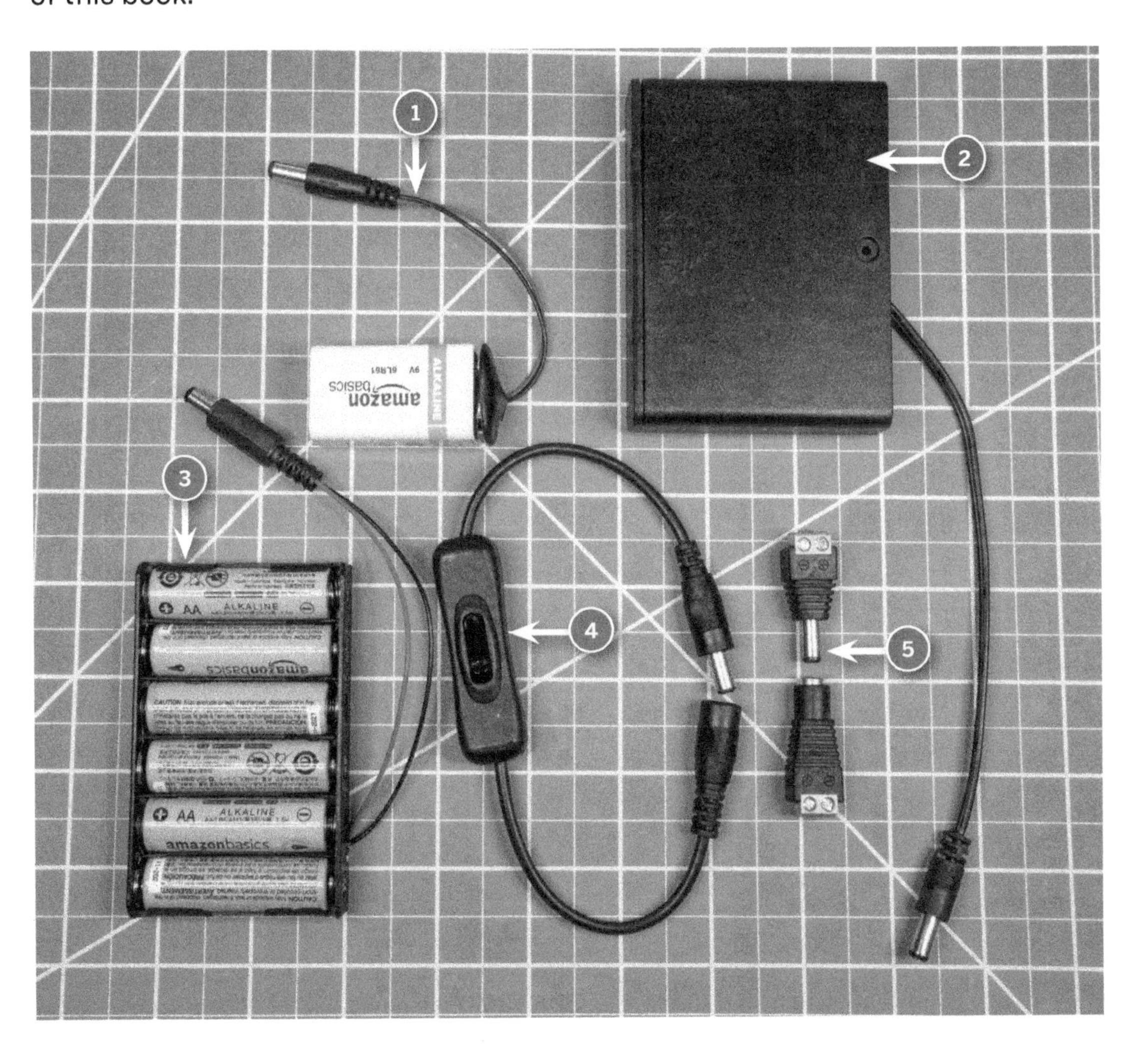

THE WIRING AND CODE

Pages 84 and 85 show the wiring and code for this project. The wiring is very similar to the last chapter except for using pin #2 on the Arduino instead of pin #3. Also, in this case, the black wire is used to ground, the red wire goes to 5V, and the white wire carries the signal. For the purposes of clarity, I used a NeoPixel graphic from the Tinkercad starters library that shows the NeoPixels arranged in a straight row instead of a circle. Functionally these are exactly the same, but I think this drawing is more straightforward. In this arrangement, the signal wire is addressing each NeoPixel separately. There is literally an address for every NeoPixel in the chain. This means the signal might say something like, "I want the third LED on this chain to turn bright pink for one second, and then I'll move on and address the fourth LED in the chain, asking it to turn a slightly different shade of pink."

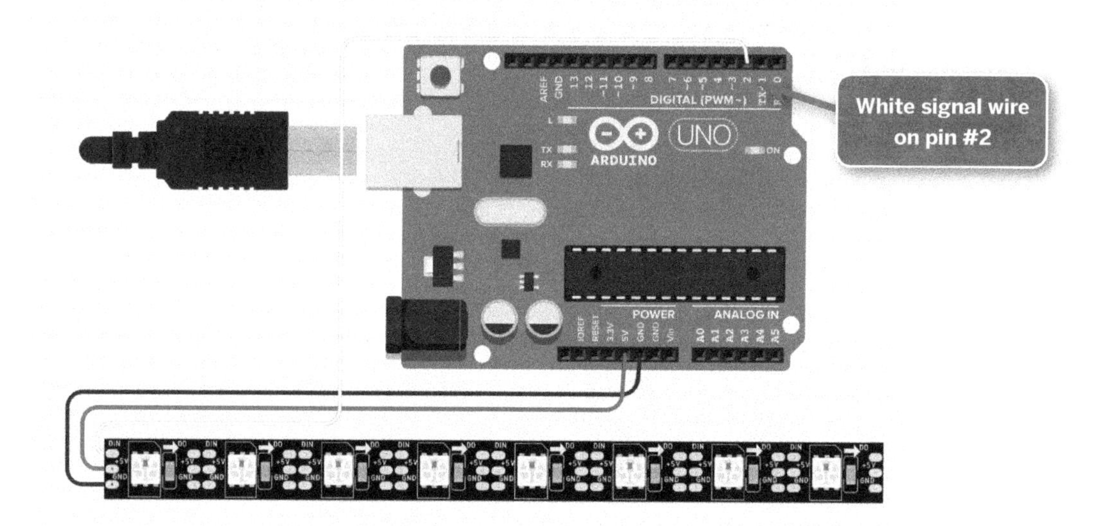

```
Text

 1 #include <Adafruit_NeoPixel.h>
 2
 3 #define PIN 2     // input pin Neopixel is attached to
 4
 5 #define NUMPIXELS      12 // number of neopixels in strip
 6
 7 Adafruit_NeoPixel pixels = Adafruit_NeoPixel(NUMPIXELS, PIN, NEO_GRB + NEO_KHZ800);
 8
 9 int delayval = 100; // timing delay in milliseconds
10
11 int redColor = 0;
12 int greenColor = 0;
13 int blueColor = 0;
14
15 void setup() {
16   // Initialize the NeoPixel library.
17   pixels.begin();
18 }
19
20 void loop() {
21   setColor();
22
23   for (int i=0; i < NUMPIXELS; i++) {
24     // pixels.Color takes RGB values, from 0,0,0 up to 255,255,255
25     pixels.setPixelColor(i, pixels.Color(redColor, greenColor, blueColor));
26
27     // This sends the updated pixel color to the hardware.
28     pixels.show();
29
30     // Delay for a period of time (in milliseconds).
31     delay(delayval);
32   }
33 }
34
35 // setColor()
36 // picks random values to set for RGB
37 void setColor(){
38   redColor = random(0, 255);
39   greenColor = random(0,255);
40   blueColor = random(0, 255);
41 }
```

HOW THE CODE WORKS

Any part of the text in Arduino code that appears to the right of the double slashes (//) is there as a human comment. The computer ignores it. The explanation for the lines of code above is as follows:

1. Line #1 is adding a library in a similar way as it did in chapter 4, when we added a library to control the servo motor. These libraries are simply underlying sets of rules that we are running in the background to make our coding job easier. In this case, we are adding a library that contains a lot of rules about working with NeoPixel strips.

2. Line #9 adds a 1/10-second delay.
3. Lines 11–13 address how much red, green, and blue (RGB) are mixed into each NeoPixel. The starting value is zero.
4. On line #20, we begin programming our loop, taking turns sending the existing brightness of each RGB color to each NeoPixel in succession.
5. Then on line #37 we exit the loop after 12 commands and set a random value for each RGB color and then turn on each of the 12 LEDs with the new color mix with a 1/10-second delay between each one.

Authors can go down a huge rabbit hole while attempting to explain why the brightness of each RGB color is set between 0 and 255. It is my suggestion that you chase that rabbit on the Internet if you want to. For now, just know that 0 is off and 255 is maximum brightness. Adding more than this number of NeoPixels for more than a short burst will wreck your Arduino.

IN THIS CHAPTER YOU:

- got a NeoPixel ring going with an Arduino.
- used a devoted external power supply to supply more current than the Arduino can provide on its own.

TAKING IT FURTHER

Do an Internet search for “light logo Arduino Josh Burker” and check out the fastest way to more complex patterns using NeoPixels. You can also expand more on the scale of your project by adding a lot of modules together. If you do this, then make sure to read up on using resistors and capacitors to safely power larger sets of NeoPixels. Adafruit is a great resource for this.

Echolocation Distance Sensor—Like a Bat!

MATERIALS USED IN CHAPTER 7:

- **HC-SR04 Distance Sensor**
- **Female-to-Male Breadboard Jumper Cables**
- **Arduino Uno R3**
- **5mm LEDs (look for an assortment box)**
- **Breadboard**

I have always been interested in how the echolocation system works for bats (the flying kind). It turns out that the reverse sensors in a car (the ones that beep as you get closer to an object behind you) work in the same way that bat echolocation does. The bat sends out a signal/pulse, then waits for it to return, and translates the elapsed time into distance. On a new car, you can see small round divots in the left and right side of the bumper. Those divots serve exactly the same purpose. They send out a signal/pulse and then the computer counts the elapsed time until the signal is returned after it bounces off a nearby object. You can buy a similar sensor for about a dollar on Amazon. It is called an "HC-SR04." If you thought the three-wire club was good, then just wait until you're initiated into the four-wire club!

The wonderful thing about the four-wire club, and every other club after that, is that your work gets no more complicated on a fundamental level. When we were using the servo, the third wire carried the rotation position signal to the servo. In this case, we are going to have one signal for the outgoing pulse and one signal to transmit the timing of the incoming pulse. Also, in the same way the signal wire shared the return trip to the Arduino on the GND, this fourth signal wire will also hitch a ride home to the Arduino via the GND wire. In the picture of the completed project shown below, if you count carefully, you will see six wires because the "taking it further" section adds LEDs that will turn red or green, depending on how close or far an object is from the distance sensor. To begin with, however, there will be a live feedback through the Arduino IDE telling exactly how far away the object is from our distance sensor in real time.

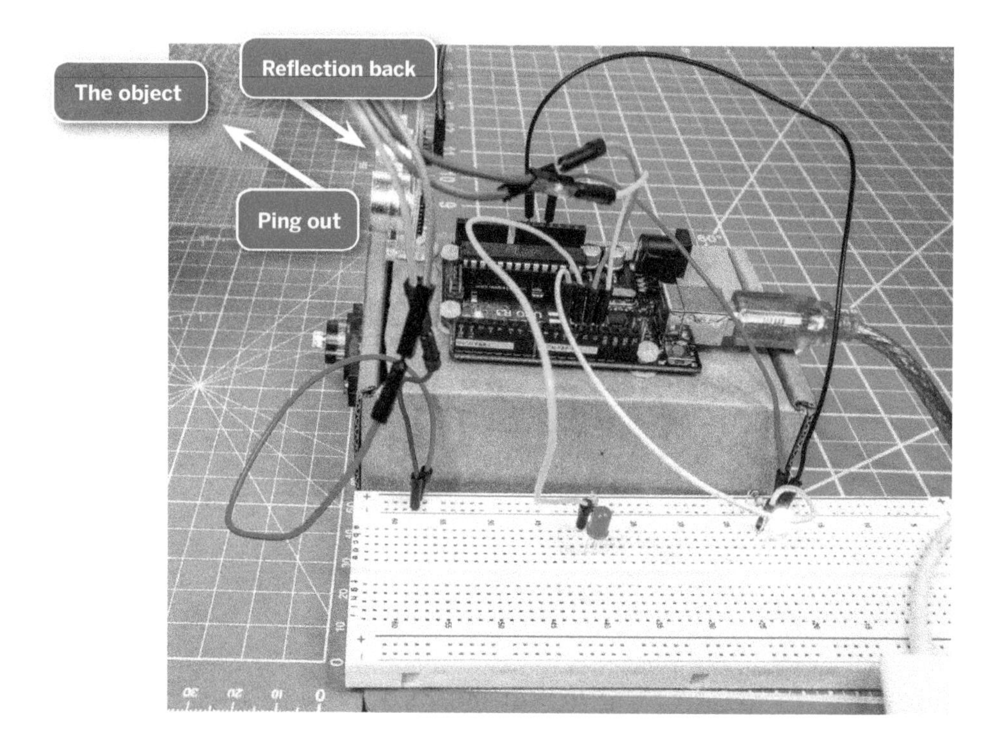

WIRING THE HC-SR04

There are a total of four wires to connect to the bottom of the HC-SR04 distance sensor. One is for power, one for ground, one for the input signal to the distance sensor, and one for the output signal from the distance sensor. The gray wire above is for the positive 5v connection. (VCC stands for "voltage common collector," which is just a complex-sounding acronym for positive wire.) The green wire on the right side is connected to the GND. The red wire in the middle left is connected to a signal wire (TRIG stands for trigger), and the blue wire on the middle right is connected to the ECHO pin.

You can mount the distant sensor in a variety of ways. The distance sensors are not affected by vibration or minor dust, dirt, and drops. Other than water, the main thing you can do to kill the accuracy of a distance sensor is to cover the front two holes. This particular distance sensor came with a plastic mounting plate, so I added a hole and a new grommet to the existing box and mounted it facing out on the corner on the servo side. You can mount it any way you choose just so long as the face is parallel with the target.

SIMULATING THE ULTRASONIC RULER IN TINKERCAD

There are a few starter examples in the circuits section of Tinkercad for the HC-SR04 distance sensor. In this example we are going to use the simplest possible model without any extra components. The wonderful thing about the simulator is that when you hit the "Start simulation" button, a green cone appears, and you can drag a ball around that area and see what readings you would get from an actual distance sensor. To see this, click on the "Serial Monitor" button when the simulation is running and see the data coming at you!

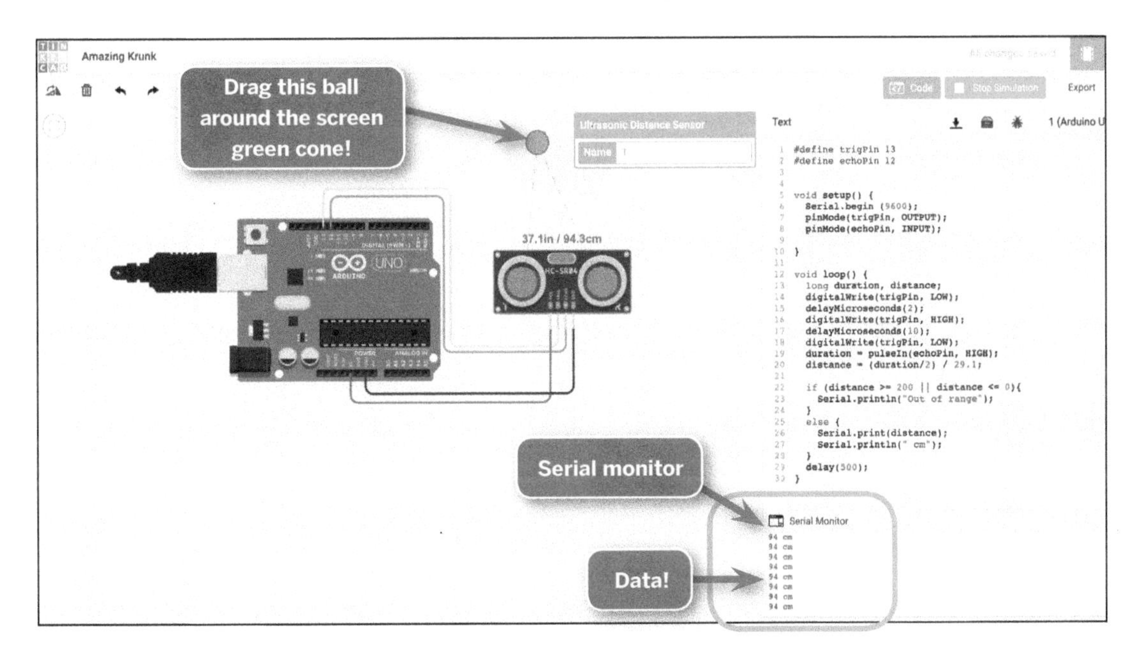

LET'S TAKE APART THE CODE

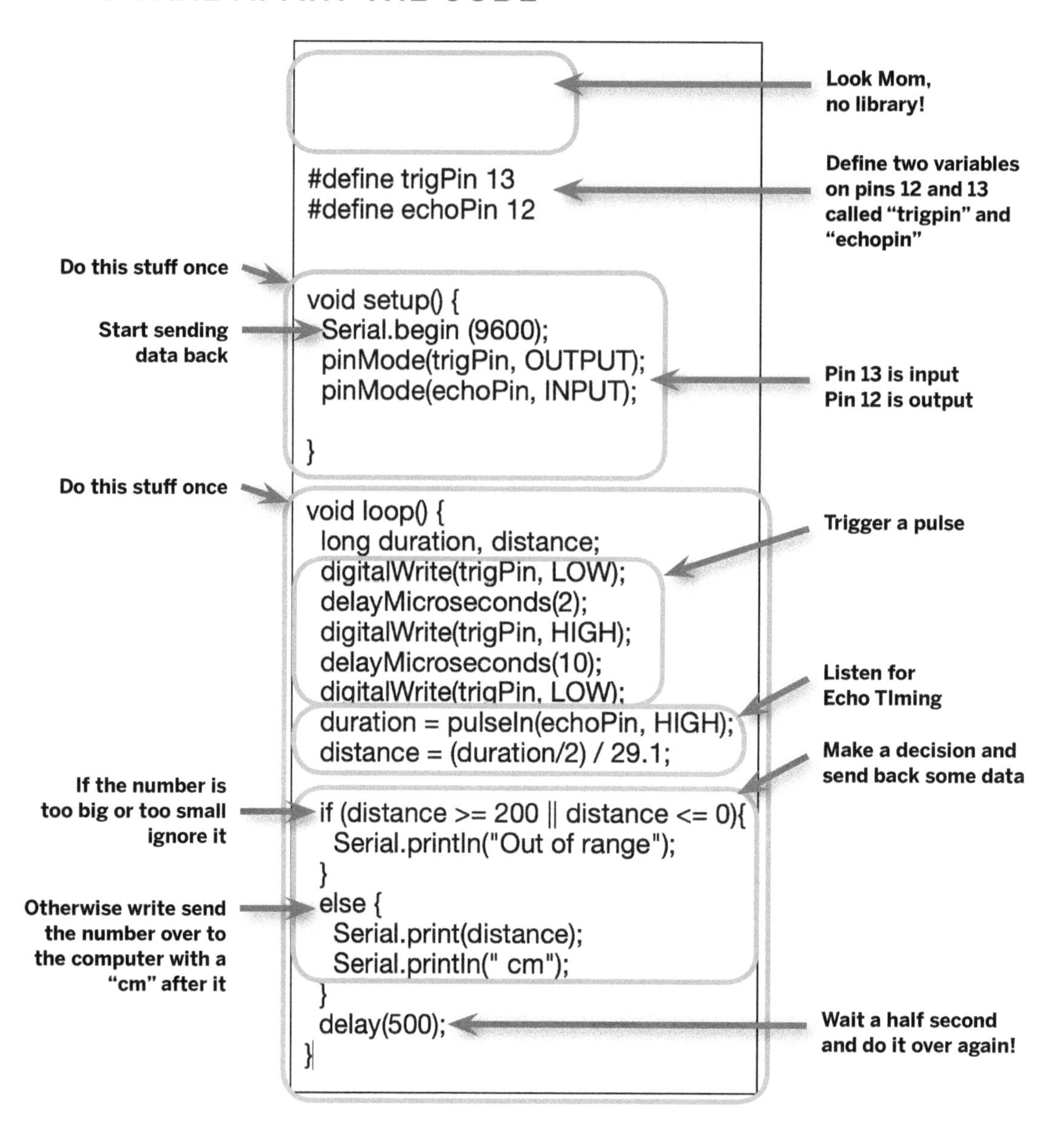

GETTING DATA

After you verify and upload the program to the Arduino board click, Tools > Serial Monitor, and a window will pop up showing you the live data coming back from the Arduino. You can choose to make decisions in your program on the Arduino without being hooked up to your laptop. In the next sketch, we will add a couple of outputs to either light up a green LED or a red LED, based on the proximity of an object to the distance sensor. This will work even when you plug in the Arduino from the computer and run it on battery power.

ADDING LEDS

Note that in the illustration below, we are adding two LEDs to the circuit. Before we go further, place your finger on the diagram and physically trace the 5V power wire from its start on the Arduino through all the number of ways it transitions into the GND (black wire) coming back to the Arduino. Then trace each of the signals we have on the board. There are a total of four signals: one for the red LED, one for the green LED, one for the TRIG signal, and one for the ECHO signal. That is a total of eight wires and six signals. A general rule is that most simple circuits we want to build will have two more wires than signals.

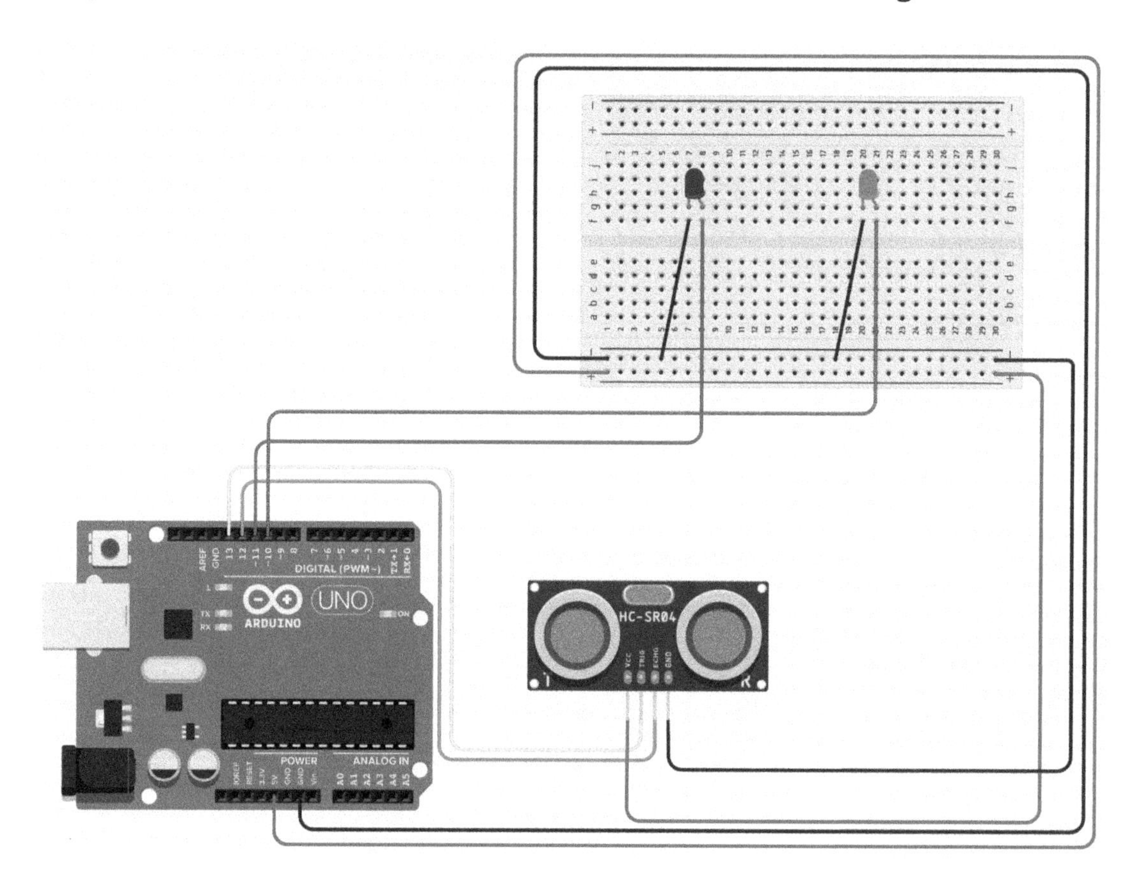

Comparing Code

What are the differences between the two sets of code? Give them a good look. Use the code on the right to complete the project by adding the two LEDs.

```
#define trigPin 13
#define echoPin 12


void setup() {
  Serial.begin (9600);
  pinMode(trigPin, OUTPUT);
  pinMode(echoPin, INPUT);

}

void loop() {
  long duration, distance;
  digitalWrite(trigPin, LOW);
  delayMicroseconds(2);
  digitalWrite(trigPin, HIGH);
  delayMicroseconds(10);
  digitalWrite(trigPin, LOW);
  duration = pulseIn(echoPin, HIGH);
  distance = (duration/2) / 29.1;

  if (distance >= 200 || distance <= 0){
    Serial.println("Out of range");
  }
  else {
    Serial.print(distance);
    Serial.println(" cm");
  }
  delay(500);
}
```

```
#define trigPin 13
#define echoPin 12
#define led 11
#define led2 10

void setup() {
  Serial.begin (9600);
  pinMode(trigPin, OUTPUT);
  pinMode(echoPin, INPUT);
  pinMode(led, OUTPUT);
  pinMode(led2, OUTPUT);
}

void loop() {
  long duration, distance;
  digitalWrite(trigPin, LOW);
  delayMicroseconds(2);
  digitalWrite(trigPin, HIGH);
  delayMicroseconds(10);
  digitalWrite(trigPin, LOW);
  duration = pulseIn(echoPin, HIGH);
  distance = (duration/2) / 29.1;
  if (distance < 4) {
    digitalWrite(led,HIGH);
  digitalWrite(led2,LOW);
}
  else {
    digitalWrite(led,LOW);
    digitalWrite(led2,HIGH);
  }
  if (distance >= 200 || distance <= 0){
    Serial.println("Out of range");
  }
  else {
    Serial.print(distance);
    Serial.println(" cm");
  }
  delay(500);
}
```

How Can I Resist?

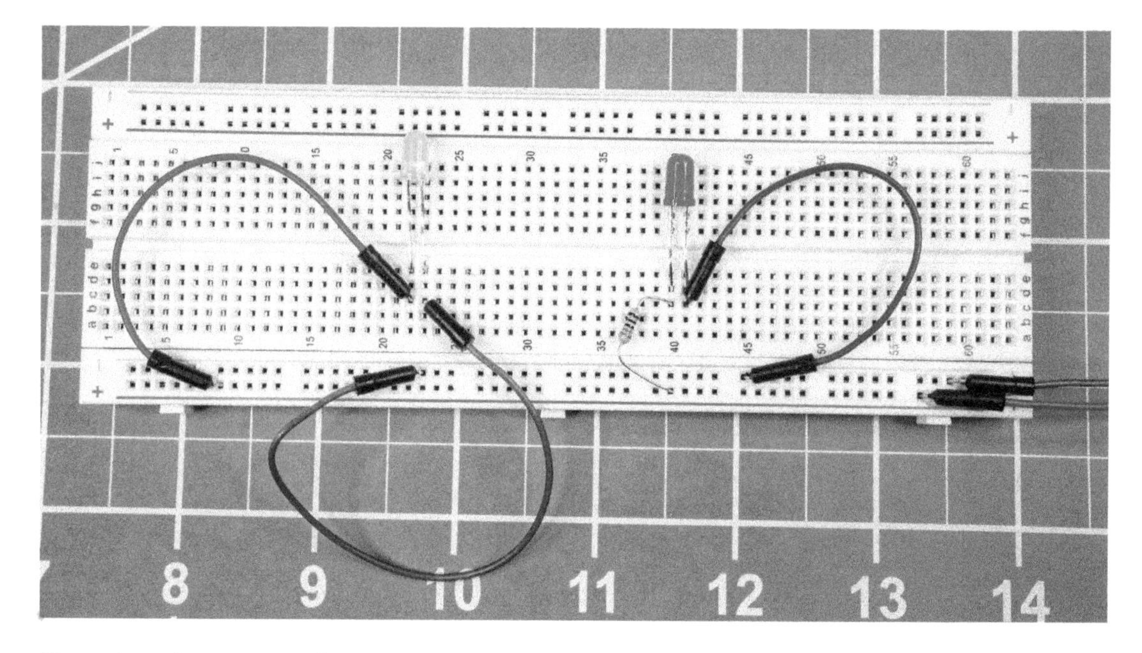

Examine the circuit above. You can see that 5V power comes into the red wire in the lower right corner and then into both LEDs. The small blue piece next to the arrow is called a resistor. There is something called Ohm's law that I have been trying to avoid sharing with you this entire book, not because the math is hard, but because electrical engineers are generally too concerned about details, and thus make things unnecessarily hard to understand. For now, just know that these resistors allow us to make small changes in the voltage that reach the LED and that within the scope of this project they are unnecessary. I bring this up in case you are growing curious from seeing resistors in the Tinkercad simulations and other tutorials.

I would suggest that you check out Ohm's law if and when you burn out a LED. Until then, I will continue resisting talking about them until we need Mr. Ohm and his laws.

IN THIS CHAPTER YOU:

- mastered an industrial ultrasonic sensor used in car backup warning systems and drone altitude systems.
- learned how sensors work and how that relates to bats in the wild.
- used a serial interface.

TAKING IT FURTHER

Can you combine the code and hardware in this chapter with chapter 3? Can you use that combined knowledge to switch on really bright lights when someone or something walks into a room? At this point you may be thinking about how each chapter can be combined with others to make more complex projects. In fact, the last chapter of this book points back to the first chapter. As you continue, keep pondering ways to combine the skills and tools in each chapter.

Big Chip and Baby Chips

MATERIALS USED IN CHAPTER 8:

- **ATTINY85 Chip w/Flat Leg Socket**
- **Index Card or Card Stock**
- **Standard Paper Hole Punch**
- **Large Paper Clip**
- **20- to 30-Gauge Wire stripper**
- **Female Breadboard Jumper Wires**
- **Small Zip Tie**
- **Sparkfun Tiny AVR Programmer**

Take a moment to pick up the basic Arduino board you have been using over the last seven chapters. Take a close look at the microchip that dominates the center of the board between the input pins and the output pins. Can you see that the chip is not soldered to the board? The chip is pressed into a socket, which is soldered to the board. Your Arduino circuit board is a well-thought-out package that houses a very common chip called the "ATmega328P-PU." This chip is found in a myriad of everyday consumer and industrial machines. Appliances, 3-D printers, and traffic automation systems all use variants of this chip. The official Arduino Uno R3 costs approximately $20. Generic copies of the board can be found for about $5. The ATmega328P-PU chip by itself retails for about $3. What if you do not need all of the features on this chip and you do not need all the components on the Arduino board? There are so many variants of this chip, some more sophisticated and some simpler. Why not choose the chip that is right for you?

In this chapter, we're going to take a huge leap forward. We are going to take a bold and brave step into the world of the professional engineer and source our own microchip for $1, and then use it to control a servo motor. Once we jump this hurdle, you will be free to try riskier and more creative projects when all that is on the line is a $1 microchip and a $1 servo motor.

We are also going to build our own circuit board! Now that you are looking at the microchip as the functional heart of the Arduino Uno, what does the rest of the board really do? To oversimplify just a bit, I would argue that the remainder of the board is a set of convenient plugs (called pins) that make it easy to plug in wires. The USB plug is also serving that same function. When we program within the Arduino IDE, we are simply sending data through a cable that eventually connects to various pins on the microchip. Everything seems to end up connected to the microchip. So let's start with requirements for our project and then select the simplest possible chip mounted on the simplest possible circuit board that will do the job that we have envisioned.

The images below are a result of a lot of exploration and a lot of collaboration with folks at Concord.org and PaperMech.net, as well as the hard work of Susan Klimczak. The chip we are focused on is called the ATTINY85, a $1 microcontroller found in many consumer and industrial products from alarm clocks to assembly-line automation. My contribution here has been around the paper egg circuit board shown in the picture on the bottom right. I encourage you to do an Internet search for "Susan Klimczak ATTINY Adventures." You will land on a page with many more details about the ATTINY85.

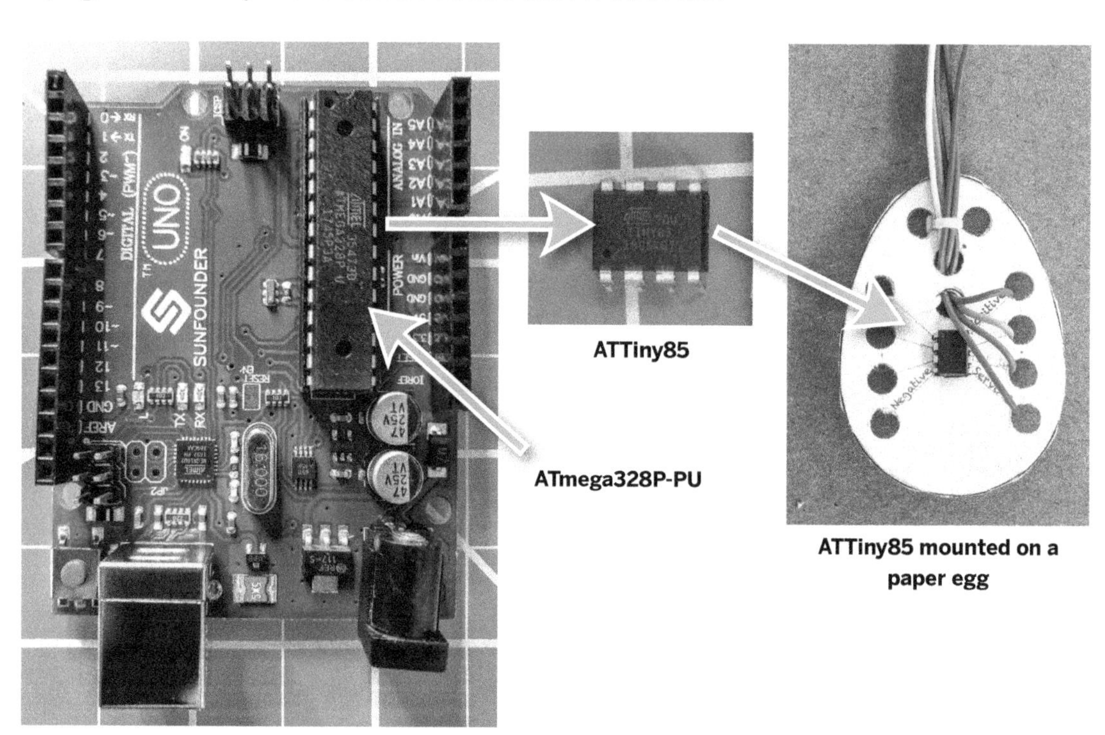

ATTiny85 mounted on a paper egg

ATTINY85 ON TINKERCAD CIRCUITS

Tinkercad for Circuits provides a great place to get started with this chip. I hope you started to explore more of the starters and examples that are available on the Tinkercad for Circuits website. This is one of those rare free and successful community websites where people from all backgrounds and interests post circuit designs and code that have worked for them. And all without ads! "Soft Servo Sweep" is one of many excellent examples on the site. The illustration below shows the circuit diagram for the wiring setup. You can use three AA batteries or simply run jumpers from the Arduino Uno's 5v and GND pins to use as a power source. Once the chip is programmed, the wiring is straightforward: just power, ground and signal wires going to pin #5.

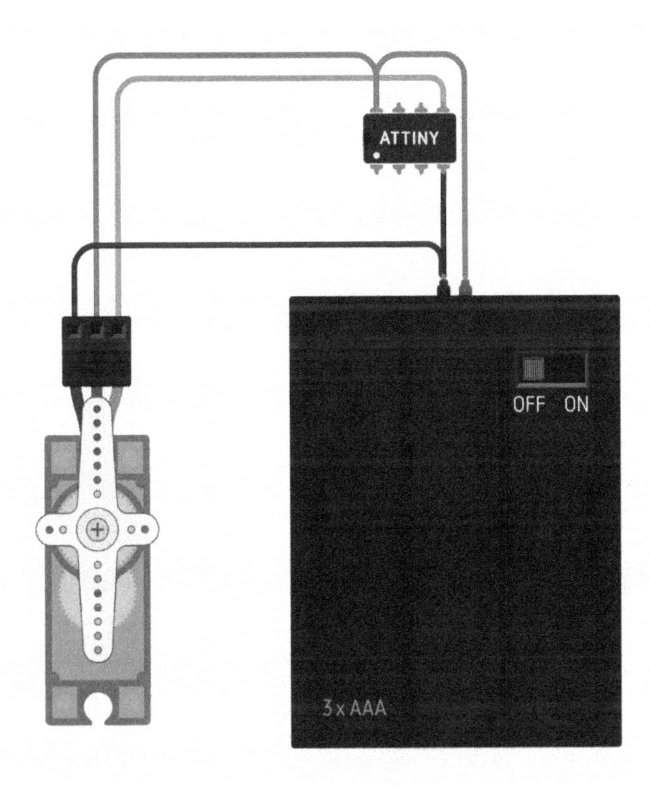

Translating the PINS

My hope is that in previous chapters you became quite comfortable hooking up various pins and relating that physical choice to our designation within software to control those physical pins. The only difference now is you are throwing away all the stuff between where we were plugging in a jumper wire on the Arduino and now connecting directly to an actual pin on a chip. You will see shortly that we will use a socket to extend the pen so we can wrap the wire around it.

Before we get to that point, though, we need to understand the layout of the physical pins on the chip. Within engineering circles this is called getting a "pinout," which is just another word for defining which pin has which number assigned to it. All chips that I know about look symmetrical and have ridiculously small writing on them which is illegible without extreme magnification. Having said that, every chip in this family (and almost every other chip that I know of) has an orientation dot. This is literally a small indent on the chip itself that is much easier to see than the faint labeling. When you fix the chip on the board, orient the chip so that this orientation dot is in the upper left-hand corner. From this orientation, the physical numbering of the pins starts at #1 in the upper left corner and proceeds counterclockwise from there. When the chip has eight total pins on it, we start with #1, then below that #2, #3, down to #4 in the lower left corner, then around to #5 in the lower right corner and up to the #6, #7, and finally #8 in the upper right-hand corner.

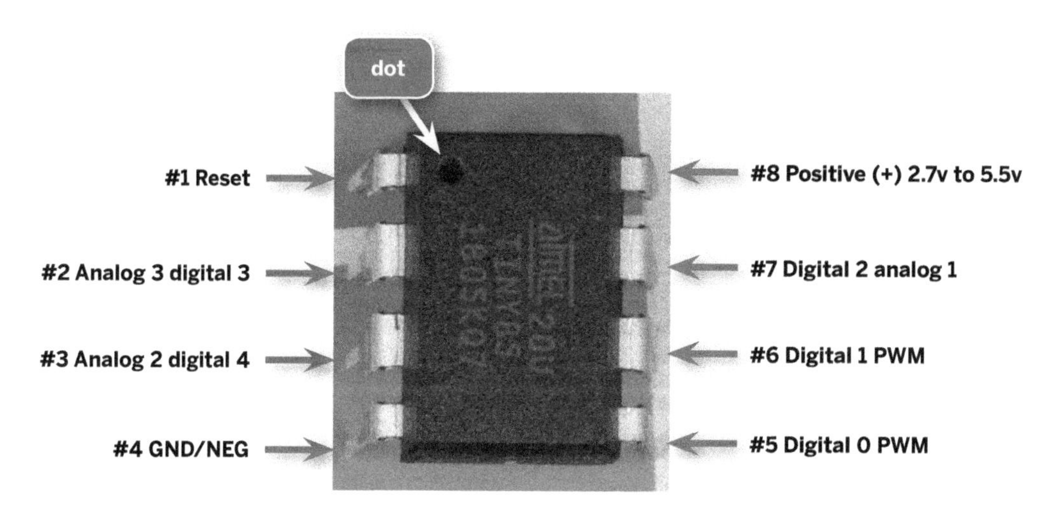

First find the power (+) and ground (-) pins and then assume everything else is a signal pin. In this case, look at #8 for power and #4 for ground. Since we are going to use a servo motor in this project, we want to find pins that have a PWM function. PWM stands for pulse width modulation, which is basically a way to encode a range of values that the servo can use to translate to a certain number of degrees. In the case of the ATTINY85, we have pins #5 and #6. The rest of the label on these two pins refer to what number our program needs to use to send a signal to the physical pin. Let's use physical pin #0. To do that, we would attach the software naming the servo motor to pin #0. There are some other very interesting signal pins on this chip. The physical pins #2 and #3 can be used for analog inputs from things like a moisture sensor or a light sensor. The physical #1 pin will reset the program every time it receives a signal.

MAKING A PAPER CIRCUIT BOARD

What we are trying to do with this paper circuit board is provide a substrate that is suitable for threading wires. The connections to and from the circuit board will be made with these jumper wires sticking up from the circuit board surface. The benefit here is due to the fact that the wires are flexible in reference to the chip. This results in a much more durable design if dropped.

Steps:

1. Use a glue stick to secure a paper design to an index card or card stock.
2. Place the ATTINY85 chip on a piece of paper.

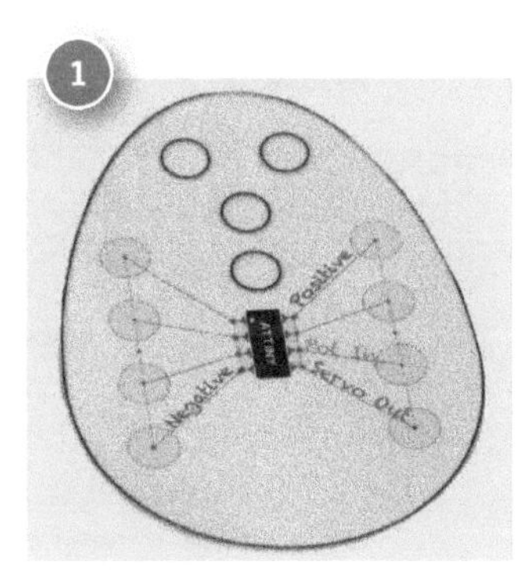

3. Lightly press the chip into the paper, leaving indentations in the paper where you can follow with a pushpin and punch holes for the legs.
4. Punch several ¼-inch holes around the perimeter and toward the center of the paper. I designed an egg shape. You can use a more creative animal or letter shape. I printed the graphics for my design in OS X Keynote. Overall the goal is to provide a semirigid surface that will be sandwiched between the ATTINY85 and the socket.

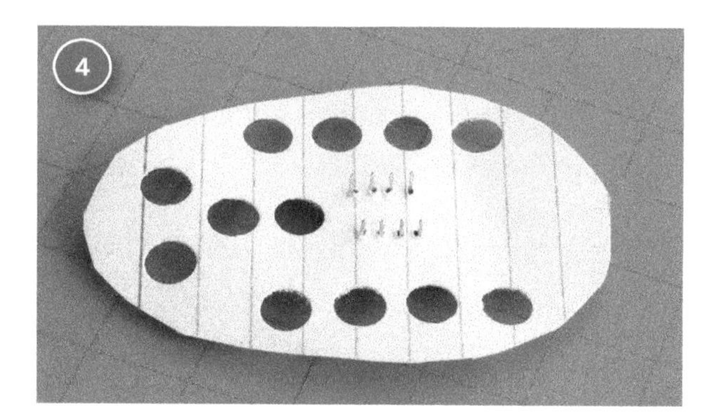

5. Once you've punched holes for the legs with a pushpin, insert the ATTINY85 chip into the paper.
6. Place the socket under the chip and push down gently. I did this by holding the socket in my left hand and holding the chip in my right hand.
7. With the test fit done, remove the chip from the socket again and proceed with wire wrapping as described on page 105.

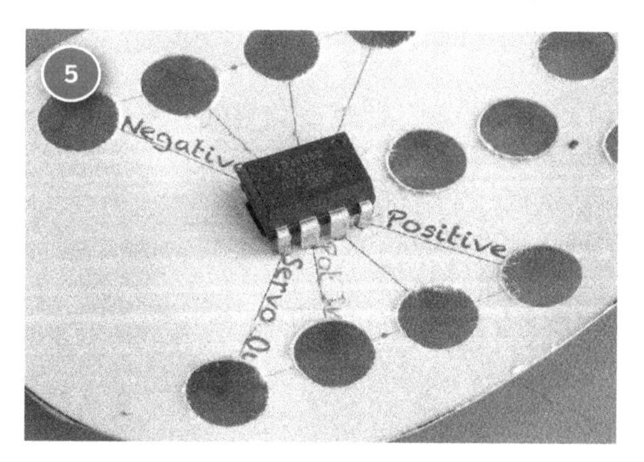

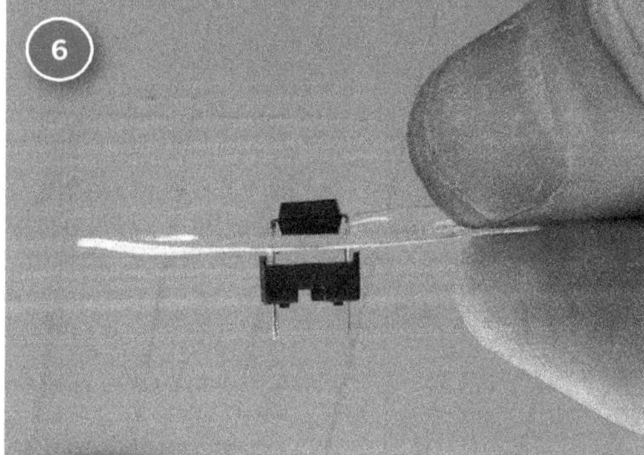

ADVANCED WIRE WRAPPING

I want to teach you wire wrapping instead of soldering because I want the projects that you build to come together as rapidly as possible so you are more confident iterating on them quickly. Soldering is a great skill, but it can slow you down. Follow the steps below to wire wrap the socket that comes with the ATTINY85 chip. (Make sure you order the chip with flat leg sockets included, as round leg sockets will not work for wire wrapping.)

Steps:

1. Use a large paper clip as a cylindrical base to bend the socket legs over toward the center. Do this for both rows of four legs.
2. Insert a wire with 1.5 inch of insulation stripped away and its tip twisted.
3. Insert that wire in the loop from the outside center bottom of the bent-over socket leg.
4. Wrap the wire back around itself a few times until it's secure.

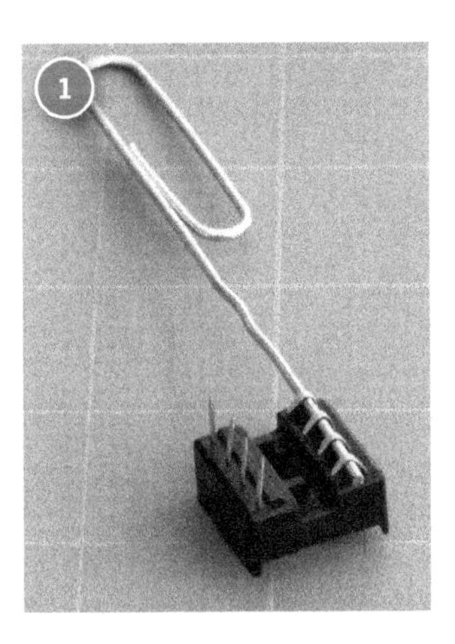

5. Trim off the excess length with scissors.
6. Repeat for socket legs corresponding to physical chip legs #8 through #4.

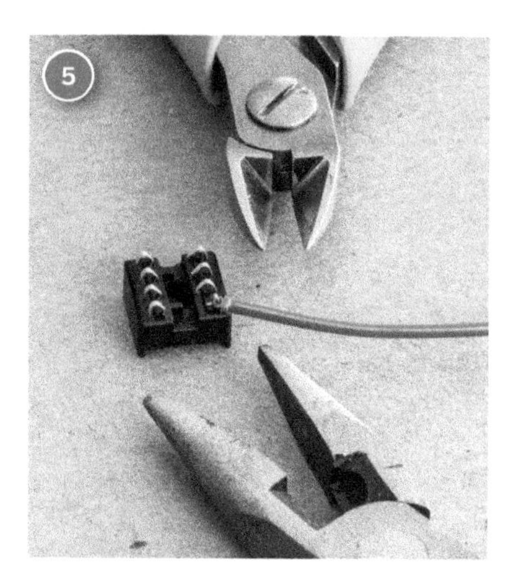

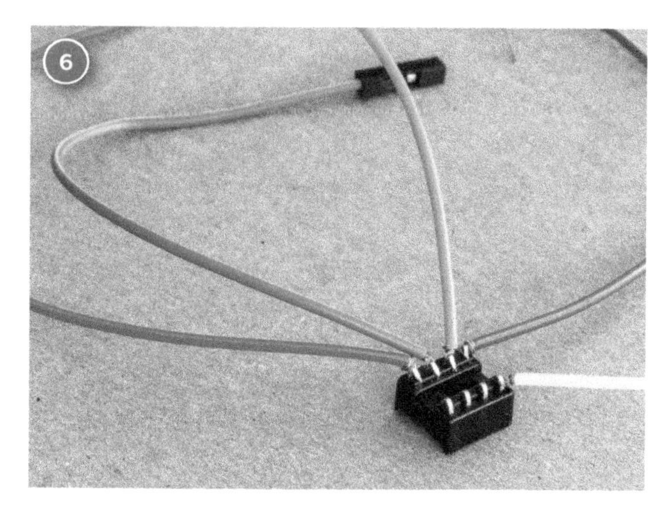

ADDING TENSION TO THE WIRE

Here's a quick and easy visual to help you understand why this wire-wrapping technique works:

1. Place both of your hands in front of you.
2. Put the tip of your thumb on the tip of the index finger on the same hand. Do this for both hands. You should be looking at two circles.
3. Open up a gap between your right thumb and index finger.
4. Then move to your left hand, closing the gap between your thumb and your index finger inside of the other circle. Now you have two interlaced circles.

This is much like the electrical connection we just made when we wrapped wire around the bent-over socket legs. If there is a small amount of room between the wire and the socket leg, you run the risk of an incomplete electrical connection. The purpose of the holes in the paper egg and wire tie on top is to permanently add pulling tension in the wire toward the outside of the socket. This is similar to your trying to separate your hands while in the position described above. There will always be contact between your left and right hand because of that outward tension. The picture below on the left shows wires dangling from the sockets ready to be threaded through the punched holes. The picture below on the right shows the completed route with the wires threaded up through the holes on the side, then down through the lower center hole, then up through the middle center hole, and finally zip-tied between the two top holes. This is all to keep tension on the wires, always gently pulling them away from the socket legs to ensure a secure connection.

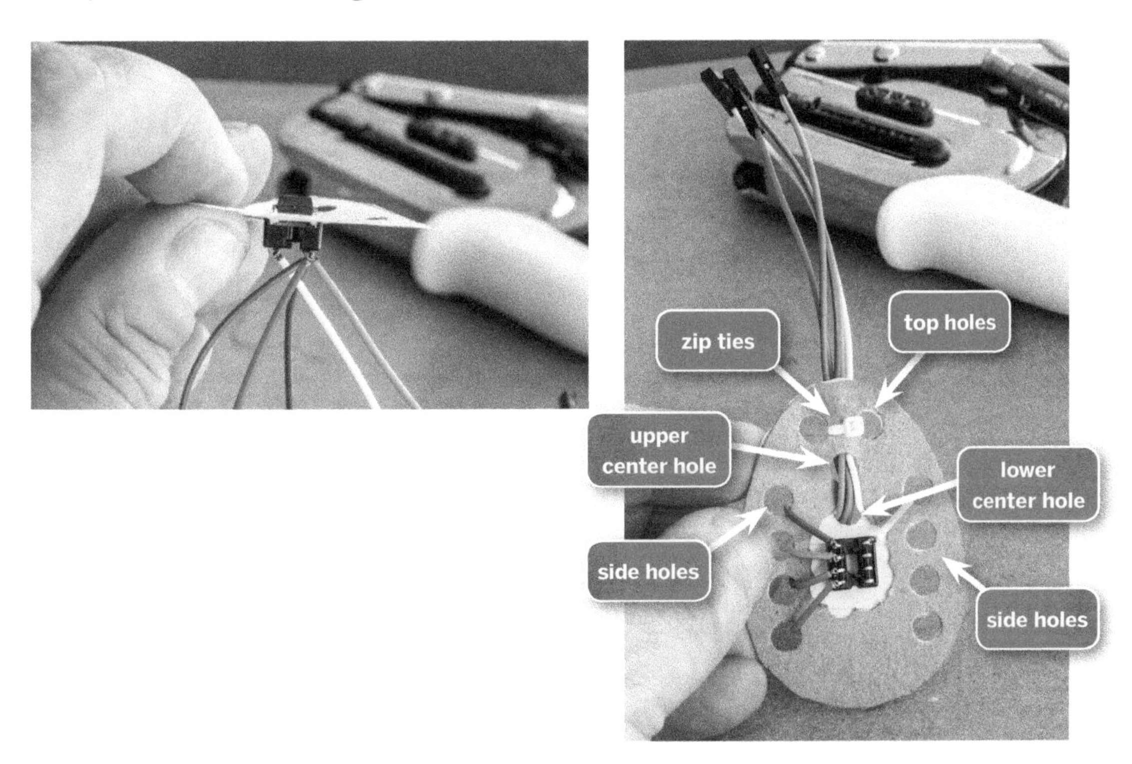

PROGRAMMING THE ATTINY85

How do you program the ATTINY85 if there is no USB port? The answer is you need a USB port. There is a more complicated way to attach the ATTINY85 to a breadboard, then hook it up to an Arduino, and then pass the data from the computer through the Arduino and into the ATTINY85. For now, we will use the

"Tiny AVR Programmer" from Sparkfun. This is a $20 USB stick where we can mount our ATTINY85 temporarily for programming. This is quite an easy process once you get through the initial setup.

Steps:

1. Take an ATTINY85 chip, locate the orientation dot, and orient it so the dot is in the upper left corner.
2. Take the Tiny AVR Programmer and orient it so the USB is on the left.
3. Locate the white line around the chip socket and look for the notch in the top.
4. Very gently squeeze all eight legs of the chip between your thumb and your forefinger, narrowing the gap between the two rows of legs by about 1mm.
5. While the chip is between your fingers, align the eight legs with the socket and push directly down.
6. Insert the USB stick into the computer before you start programming. (Caution: inserting the USB stick after you start changing settings within the Arduino IDE will result in the computer failing to recognize the USB stick.)

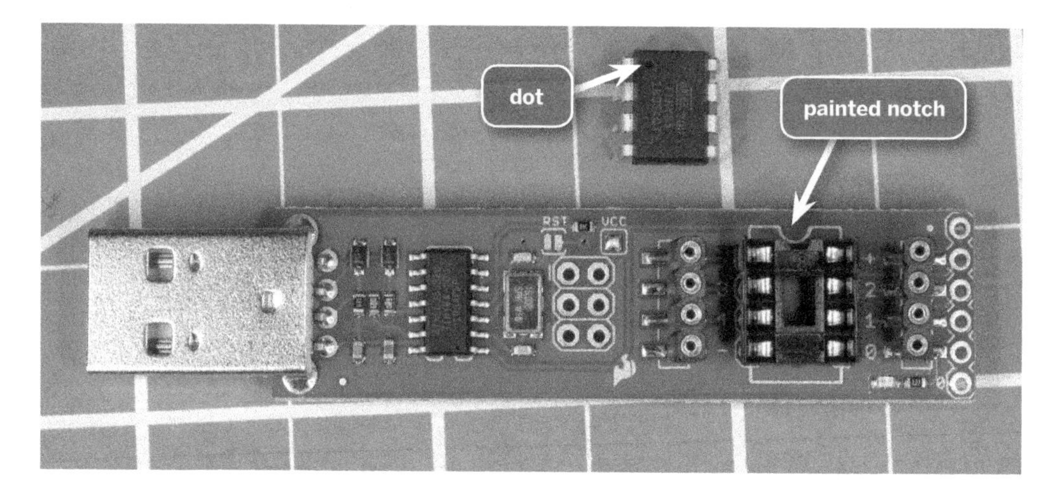

GETTING THE ARDUINO IDE TO WORK WITH THE ATTINY85

Below are the steps to get the ATTINY85 working with the Arduino IDE.

Steps:

1. Plug in the ATTINY85 to the Tiny AVR Programmer.
2. Add the chip in the Tools > Boards > Board Manager > Search ATTINY. Be sure you plug your ATTINY.
3. Choose Tools > Board > ATTtiny85 (8MHz internal clock).
4. Choose Tools > Programmer > USBTinyISP.
5. Choose Tools > Burn Bootloader. Note: this needs to be done every time you plug in a new ATTINY85 that you have never programmed before.

These steps only need to be done once.

The manufacturer of the Tiny AVR Programmer has the best instructions for the Windows operating system here: https://learn.sparkfun.com/tutorials/tiny-avr-programmer-hookup-guide/all.

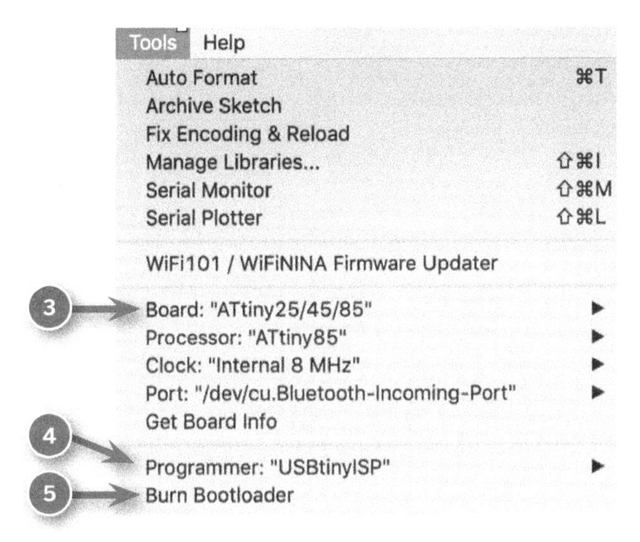

UPLOADING CODE TO THE ATTINY85

The screenshot below shows a simple test program that will blink the LED of the ATTINY85 every half second while it is still on the Tiny AVR USB stick. It will start blinking as soon as the program is uploaded. Proceed to Tinkercad Circuits for other designs like the Smooth Servo Sweep. Search for ATTINY85.

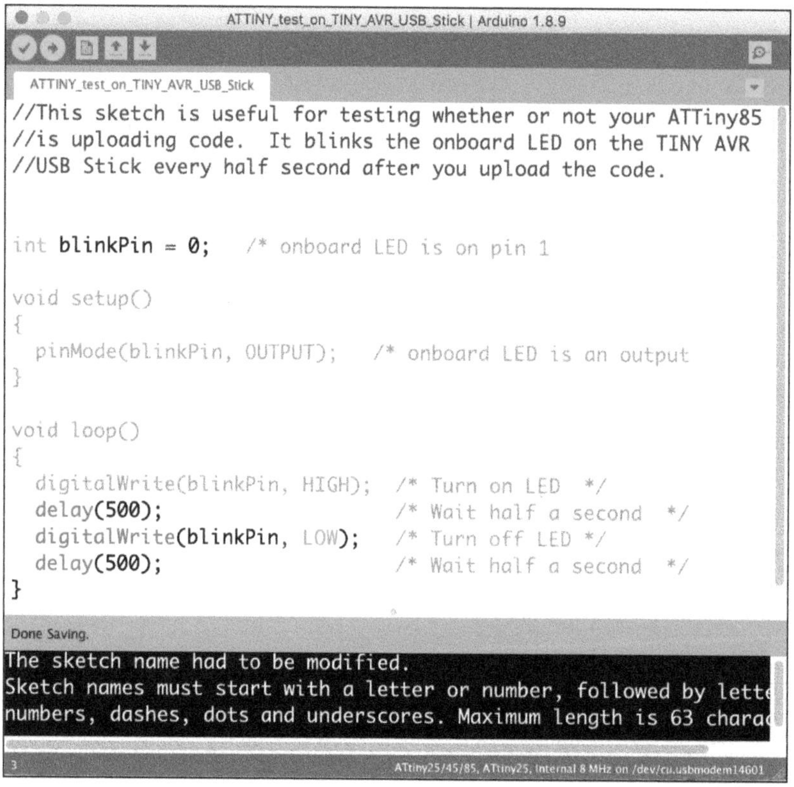

ATTINY_test_on_TINY_AVR_USB_Stick | Arduino 1.8.9

ATTINY_test_on_TINY_AVR_USB_Stick

```
//This sketch is useful for testing whether or not your ATTiny85
//is uploading code.  It blinks the onboard LED on the TINY AVR
//USB Stick every half second after you upload the code.


int blinkPin = 0;    /* onboard LED is on pin 1

void setup()
{
  pinMode(blinkPin, OUTPUT);    /* onboard LED is an output
}

void loop()
{
  digitalWrite(blinkPin, HIGH);  /* Turn on LED  */
  delay(500);                    /* Wait half a second  */
  digitalWrite(blinkPin, LOW);   /* Turn off LED */
  delay(500);                    /* Wait half a second  */
}
```

Done Saving.

```
The sketch name had to be modified.
Sketch names must start with a letter or number, followed by lett
numbers, dashes, dots and underscores. Maximum length is 63 chara
```

3 ATtiny25/45/85, ATtiny25, Internal 8 MHz on /dev/cu.usbmodem14601

IN THIS CHAPTER YOU:

- used the Arduino interface to program an industry standard chip. The Arduino Uno R3 is an accessible board for education. The ATTINY85 is a very inexpensive microcontroller used by engineers in a wide variety of products you see every day. You are now one of those engineers!
- used an index card or card stock as a substrate for a circuit.
- connected the pins on a microchip without solder.

TAKING IT FURTHER

There is a whole family in the AT line of chips. Check concord.org and PaperMech.net to seek out more projects with the ATTINY85. With this smaller form factor, you can pack a lot of project into a very tiny package!

Make a Crazy Clock with Stepper Motors

MATERIALS USED IN CHAPTER 9:

- **ULN2003 Stepper Motor Controller**
- **28BYJ-48 Stepper Motor**
- **Clockface and Clock Hands**
- **Arduino Uno R3**
- **Breadboard Jumper Wires**
- **6 x AA battery box with 2.1mm x 5.5mm Barrel Connector**

An Arduino can control both servo motors and stepper motors, ensuring that they rotate to a specific angle and move at a specific speed.

Servo motors can rotate from 0 to 180 degrees, and they "know" where they are thanks to a sensor called a potentiometer that sends a feedback signal to confirm the angle. This means that a servo motor can increase the amount of power it uses to hold a certain angle. This also means that the harder you push a servomotor, the more it will push back until it breaks a gear and fails. This is very different from a stepper motor with its unlimited 360-degree rotation that has no feedback circuit. This means that a stepper motor can be overcome with too much input torque (load on the motor), but when it fails to hold position, it will simply rotate in the direction of the greater force without permanent damage.

From a practical application standpoint, a stepper motor would make a great clock if you were to add hands to it, while a servo would work great to wave a flag back and forth.

The next item necessary for this project is referred to as ULN2003, and it is another $1 component. This is the control board (green on page 114) that translates the signals sent from the Arduino and amplifies them to the point that they are strong enough to energize electromagnets inside the 28BYJ-48 stepper motor (which also costs approximately $1 on Amazon). We will be sending signals from Arduino to the control board, and the board will send those to the stepper motor.

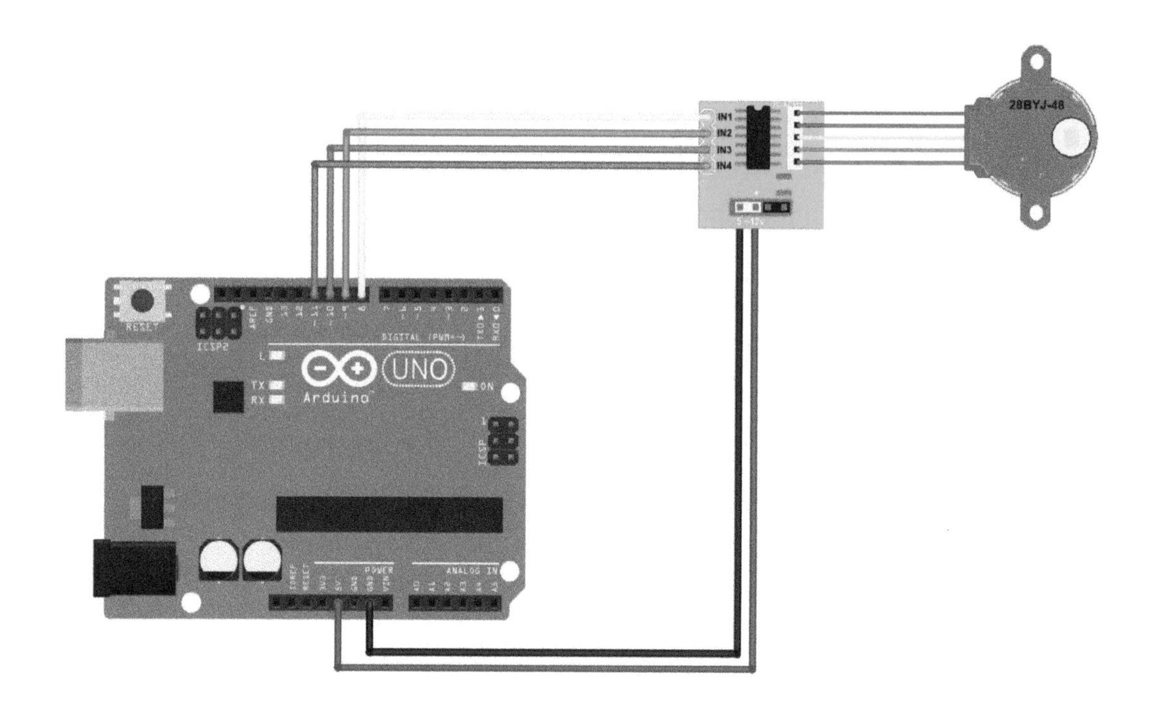

HOW STEPPER MOTORS WORK

Inside the stepper motor are coils that form different electromagnets. Look at the image on page 115 and picture someone holding a magnet at each of the north, south, east, and west positions. Imagine that the blue needle is free to pivot around the orange circle in the middle. It is attracted to the strongest magnet.

In case #1, the person is standing in the north with the strong electromagnet, and the three people standing west, south, and east have their magnets turned off. Therefore the needle points due north.

In case #2, the person standing in the north has NOT turned off the magnet, but the person standing in the east has turned on his magnet at the same strength as a person in the north. This results in a tug-of-war between the north and the east, and the needle ends up exactly between the two at 45 degrees off north.

In case #3, the person in the north is still holding her magnet at full strength, but the person in the east has turned his magnet down to half power. Therefore, the needle lands at 22.5 degrees (which is half of 45 degrees).

For case #4, can you decide the configuration of the magnets in the north, east, west, and south?

Once you understand that we are going to have four signals coming into the motor control board off the Arduino and then those signals will be amplified and sent into the stepper motor, we are ready to move on.

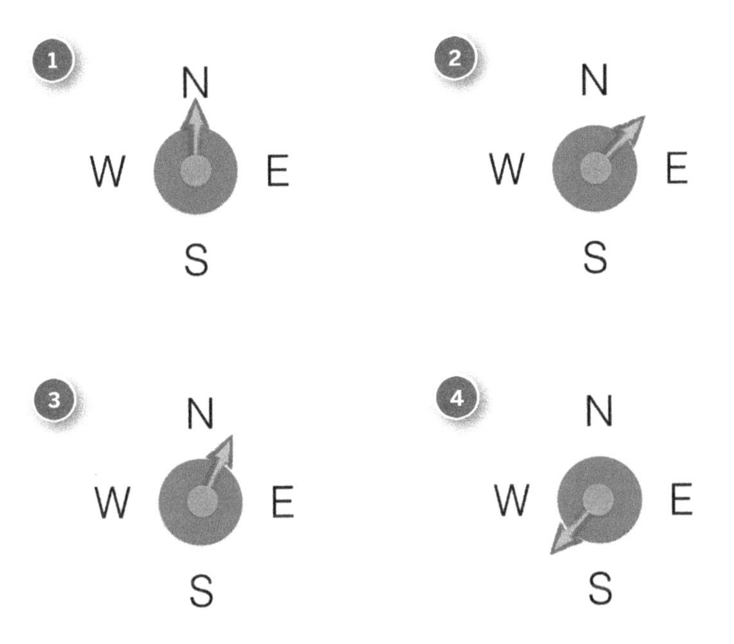

ROTATING

Page 116 shows a simple program that will allow you to get started with the stepper motor as quickly as possible. You can see that there is a library included in this sketch, which means that we will need to make sure the library is installed and accessible to the Arduino IDE. However, I suggest that you simply build and run the program and then verify it without taking the time to try to install a library. The reason for this is the newer versions of the Arduino IDE are including more and more libraries by default. <Stepper.h> is an increasingly common library, and my guess is that you have a reasonable chance of finding it already installed on your current version of Arduino IDE. If that is not the case, follow the same steps that you used before with the servo motor to install the <Stepper.h> library.

Looking at the code below, you can see that:

1. This is the library that we're calling on.
2. This is an interesting one: as long as those people standing on the north, south, east, and west turn their magnets on and off in the correct order, the needle will rotate around the circle. But how do we know how many pulses sent to the stepper motor equals one revolution of the motor? This is a place in the code where we tell the Arduino that it takes 2,038 pulses to equal one rotation of the motor. This number changes quite a bit with different stepper motor designs. This is an essential number to have in order to make your code get the expected results with your project.
3. We don't need to put anything in the setup area in this case.
4. These are the pin numbers that we are going to use to send signals from the Arduino to the motor control board. Order matters here.
5. Here we set the speed in revolutions per minute.
6. Wait a second before reversing direction.
7. Go back in the opposite direction one rotation.

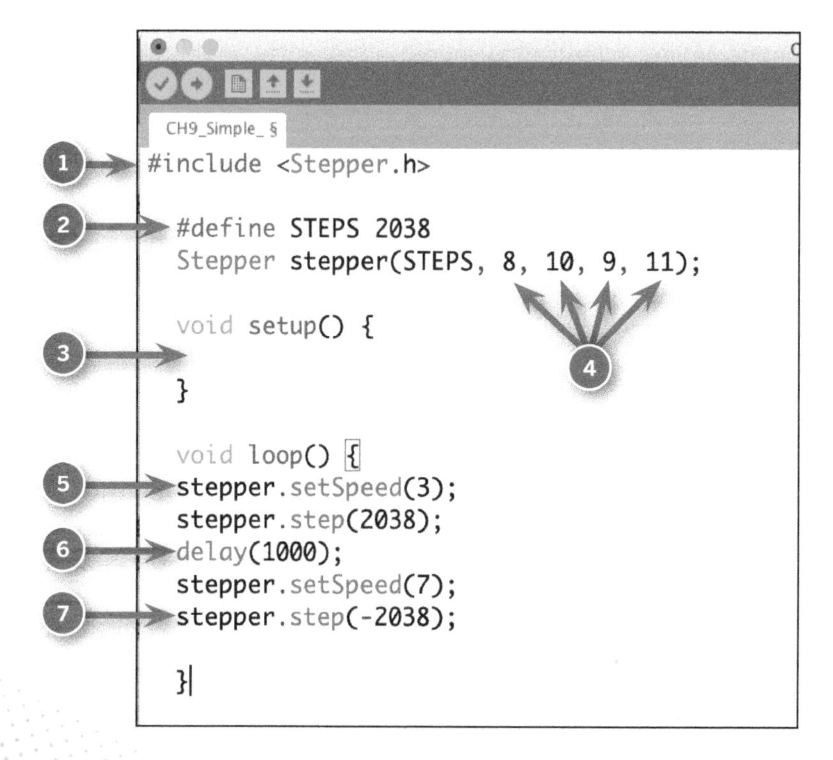

MAKING A CRAZY CLOCK

The image below shows the motor control board with wires on the left coming from the Arduino and wires on the right going to the Stepper motor. Note that wire IN1 corresponds to pin 8 on Arduino, IN2 to pin 9, In3 to pin 10, and IN4 to pin 11.

The image below is a close-up of the reverse view coming from the Arduino to the motor control board. Once again wire IN1 corresponds to pin 8 on Arduino, IN2 to pin 9, In3 to pin 10, and IN4 to pin 11.

As you can see, I kept adding projects to the same cardboard box chassis. I really have no idea what this creature will become, but it'll be interesting to see it evolve and find its purpose in life.

IN THIS CHAPTER YOU:

- used a stepper motor to achieve hyper accurate positions and rotational speeds not possible with a standard motor or servo.
- discovered that a stepper motor is a great option for continuous rotation and selecting the same spots on a dial again in a repeated pattern.
- used exactly the same hardware category that is used to build 3-D printers from scratch.

TAKING IT FURTHER

The next step is to move much larger things with larger stepper motors. The same exact principle (and the same code) works on larger motors. The industry standard is called "Nema 17," which refers to the size of the motor faceplate. Adafruit is a great place to get started with a Nema 17 motor and the Adafruit Motor Shield V2. This combination allows you to use existing pins on the Arduino and an external power supply. If you don't want to solder and are willing to use an older, but durable, motor driver, search out "L298N" on Amazon.

Key Fob, House Fob?

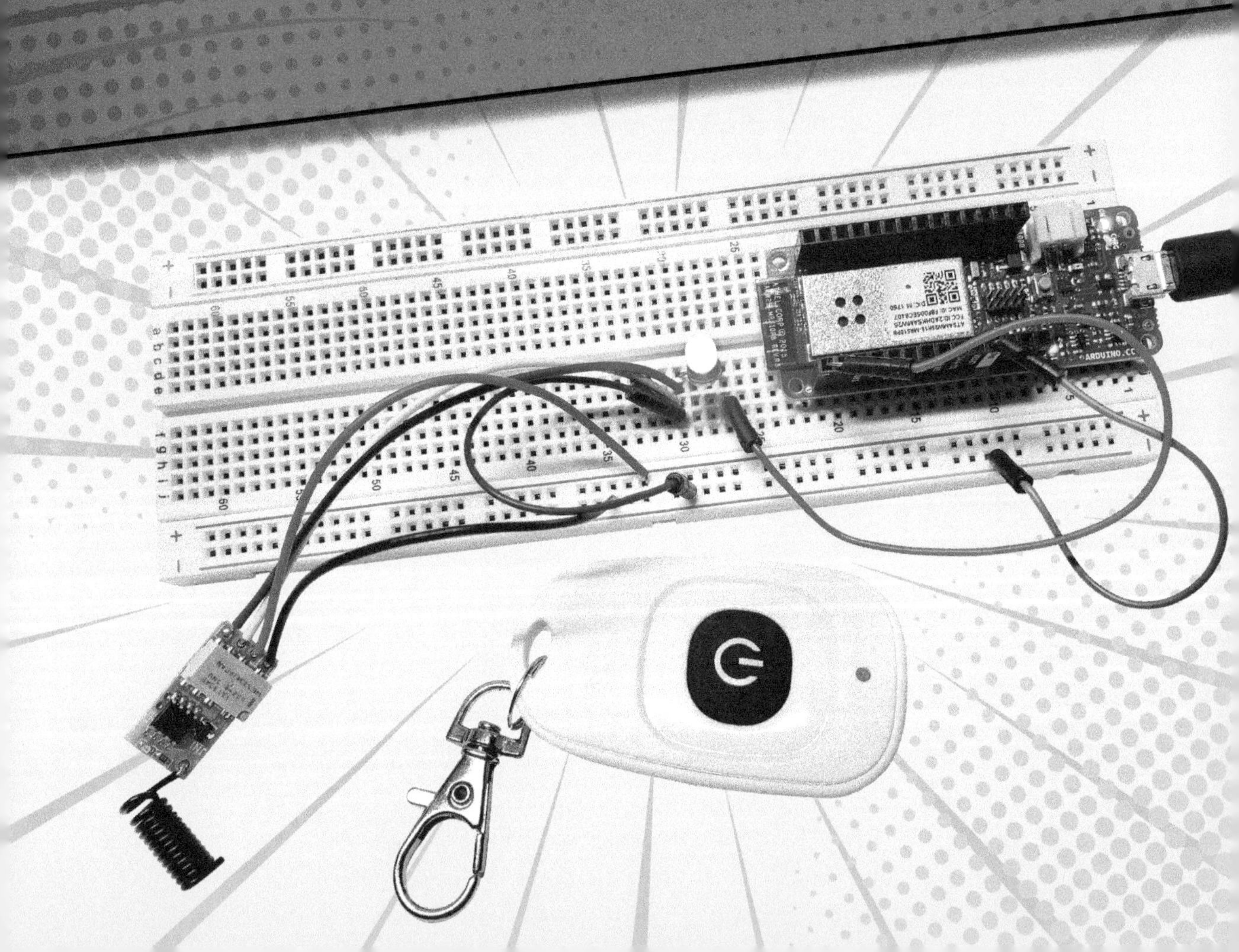

MATERIALS USED IN CHAPTER 10:

- **MKR1000 Arduino Board**
- **10k Ohm Resistor**
- **Wireless Remote Relay 12V, Momentary Switch 433mhz, DC 3.7-12v**
- **Breadboard Momentary Push Button**
- **5mm LED**
- **Male-to-Male Breadboard Jumper Wires**
- **Breadboard**

While building the project in this chapter you will realize that controlling all the lamps in the room with a wireless key fob is really, really easy. You will then recall near the beginning of the book how you used an Arduino to control a large relay with a small signal wire. What can you do when the signal wire becomes wireless? Even more exciting than that, you will realize somewhere in the next chapter that the wireless signal can actually be triggered from a web page by anyone on your network.

In the last chapter you realized that the Arduino Uno is simply an affordable and well-designed package for a microcontroller chip and that the chip is one of many generic and almost free chips available to you. The board is optional. So let's pick up a different package for a different chip, also designed by Arduino, called the MKR1000. The chip on this board is packaged with another chip that allows a Wi-Fi connection. Just like the computers on your home network, this has an IP address and MAC ID. We will explore that feature in the next chapter, but for now let's do two things:

1. Open an example sketch in Arduino called "Button" (TUTORIALS > Built-In Examples > 02.Digital > Button).
2. Explore the need for the resistor, without getting too deeply into math.

The image on page 122 shows the MKR1000 on a breadboard. Take a moment to look at the pin layout.

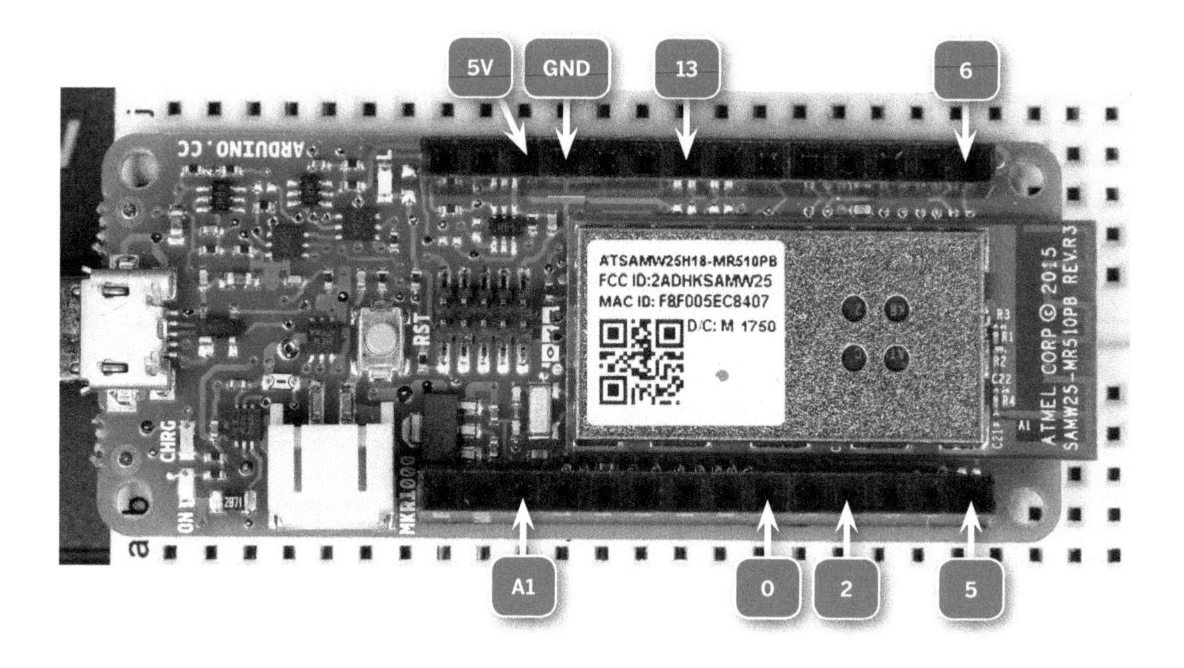

ADD THE BOARD TO ARDUINO IDE

Add the board via the board manager and then select the board. For this chapter we will program the board with a USB to micro USB cable.

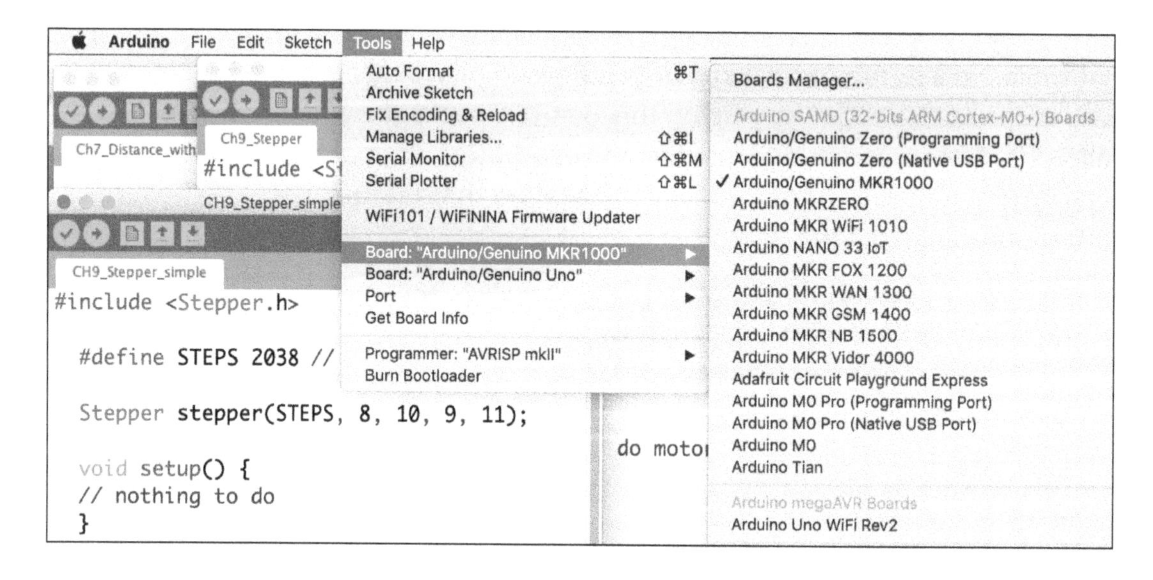

Even if you are using an OS X or Windows machine, open the example in the Arduino online interface or reference the screenshot below as we consider the wiring on the circuit.

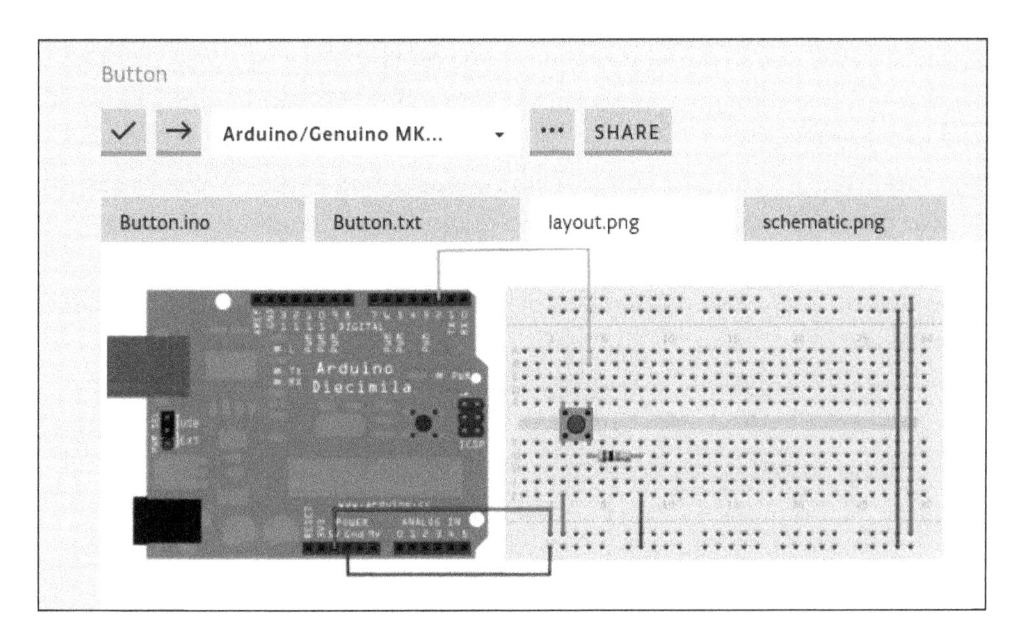

THE CIRCUIT

With all the wonderful documentation that comes with the Arduino IDE, there is some context missing that I will explain here. We are comfortable with the idea of an electrical switch. It is the equivalent of cutting a wire in a circuit or reconnecting it. In the button example, they are presenting a very different approach. In this case the microcontroller is always checking to see if pin #2 is receiving electricity. This state is called HIGH. The pin is considered to always be HIGH (on) or LOW (off). It is acting as a sensor or an input. If this is the case, the program will then activate pin #13, which is also connected to the built-in LED on the board. That is how the circuit and program differ functionally from a regular switch.

Why is this so complicated? This is the building block of digital control on which our modern world is based. If you have flown in a modern jet or driven a car made in the last decade, then you are relying on digital control. When the pilot pushes or pulls on the control stick on the airplane, it is not connected directly to the control surfaces on the wing. A digital signal is sent to a computer that the sends a signal to a powerful hydraulic actuator which moves the control surfaces on the wing. This allows the pilot to control a plane that weighs approximately 87 tons with minimal physical effort. The microcontroller is running some code to check if a pin gets a signal, and when the pin gets that signal it then turns on another pin.

When you are driving a modern car, the same system is in place. You press on the accelerator, but it is not connected to the engine. The accelerator is connected to a computer. When the computer sees you press the accelerator, it sends a signal to the engine to go faster. This oft-used program should be called: "The Modern World We Live in for Better or for Worse."

I will admit that earlier when I was dismissive of resistors, I was hoping you would be curious and do some research about them. I will now address the need for resistors. If we connect a pin directly to negative/GND, it is like trying to fill up a bucket with no bottom. The water just pours through. If we add a resistor, it is like having a bucket that has a bottom with a specified-sized hole. A large hole would be a resistor with a small Ohm value (allowing a lot of water to flow), and a small hole would be a resistor with a large Ohm value (very little water flows out of the bottom). In this case, we are using a 10k resistor, which is like having a medium-sized hole in the bucket. In the diagram below, the switch does not allow electricity to flow from left to right. It always conducts electricity up and down. When you push the button, electricity is still allowed to flow up and down but can now also flow from left to right. This means pin 2 is normally "draining" to GND unless the button is pressed, in which case 5V flows into pin 2. When that happens, pin 2 is still "draining," but the 5V flowing in is like quickly dumping another bucket of water into the pin 2 bucket so it stays full on "HIGH"—even though it is leaking—until you turn off the deluge of new water. This concept is called a "Pull-Down" resistor.

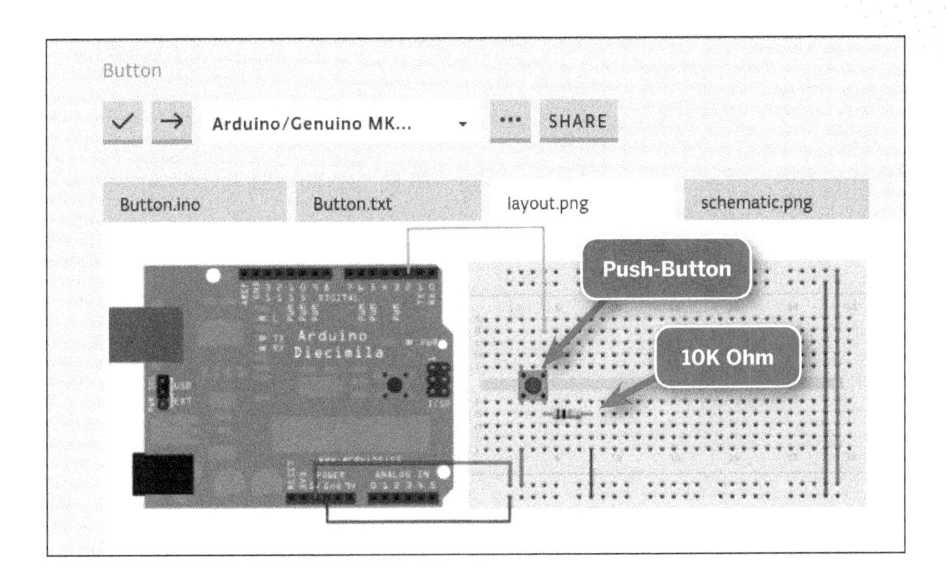

Build the Circuit

Build the circuit shown below. I have added a blue LED coming out of pin 13 so it is easier to see than the built-in LED. Even if you feel confident that you completely understand the concept of a pull-down resistor, I encourage you to place your finger on the picture and trace out the connections from 5V, GND, pin 2, and pin 13 to really be sure you see what is happening before you continue.

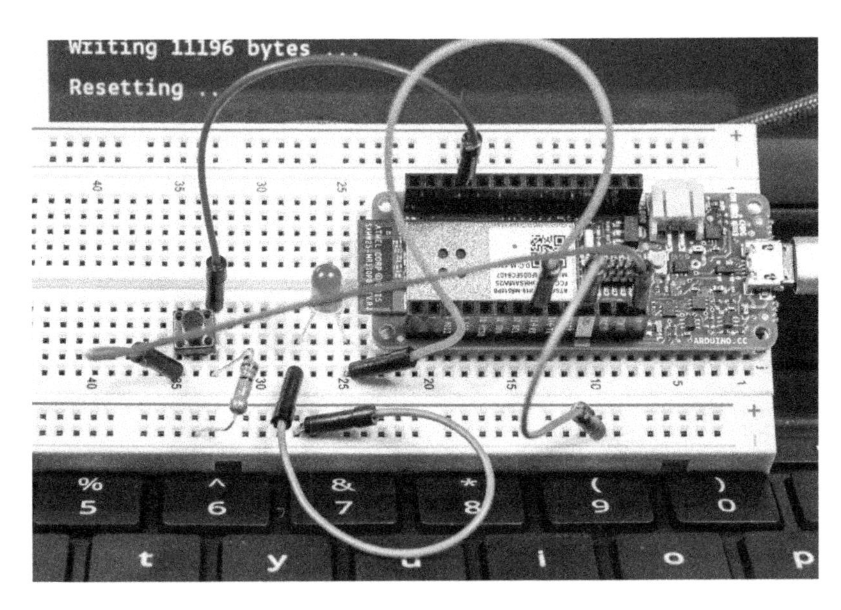

Run the Code

I have trimmed down the code from the Arduino example shown below. The full example has a lot of great comments, but for now I just want you to focus on getting this working so we can substitute the push button switch for the wireless relay key fob. One new note: "const" means "constant," or a value that will not change like a variable. A variable can stay constant, but a constant cannot vary. Try the code first as presented, then delete the "const" in the first two lines and run it again to see what happens.

```
const int buttonPin = 2;
const int ledPin =  13;

int buttonState = 0;

void setup() {
  // initialize the LED pin as an output:
  pinMode(ledPin, OUTPUT);
  // initialize the pushbutton pin as an input:
  pinMode(buttonPin, INPUT);
}

void loop() {

  buttonState = digitalRead(buttonPin);

  if (buttonState == HIGH) {
    // turn LED on:
    digitalWrite(ledPin, HIGH);
  } else {
    // turn LED off:
    digitalWrite(ledPin, LOW);
  }
}
```

Swap Switch for the Wireless Relay

When the sketch is working with the switch, remove the switch and relay. It would be a good idea to power down while doing this, but is not dangerous if you choose not to. It is wise to get in the habit of powering down a circuit so you won't get hurt when you work with higher-voltage circuits.

KEY FOB

We will use a wireless relay that reacts to the button push on a key fob. These cost about $12 on Amazon (search for Wireless Remote Relay 12V, Momentary Switch 433mhz, DC3.7V-12V). There are four wires coming out of the receiver chip. The red and black wires are the power and ground and should be hooked up to the 5V and GND pins on the Arduino. The other two wires, in this case yellow and blue, are the actual switch. These are the wires you will put into the same place that the push button switch was located. Pull out the push button switch and replace it with the yellow and blue wires. In this case, unless you are pressing the button on the key fob, electricity will not flow from 5V to pin 2 (left to right), where the switch was. Only by pressing the button on the key fob does this wireless relay allow the flow of electricity from the yellow to blue wire.

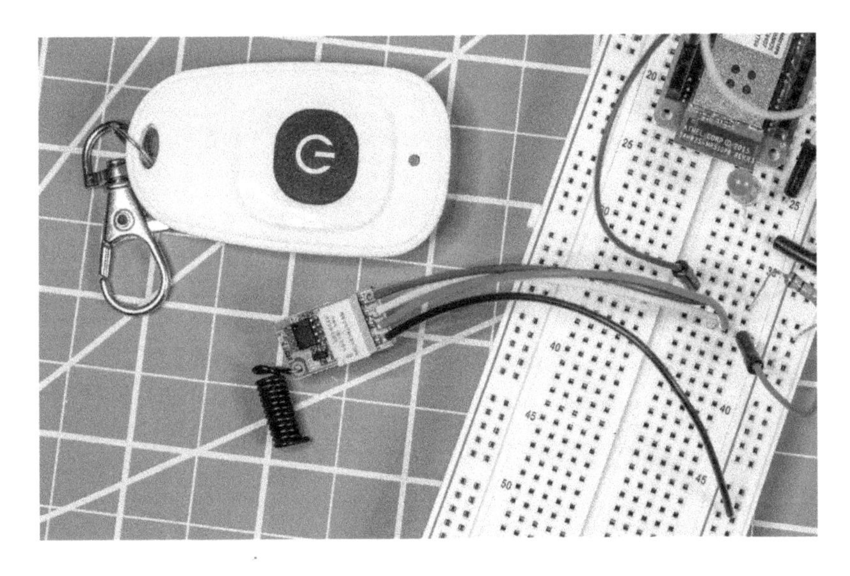

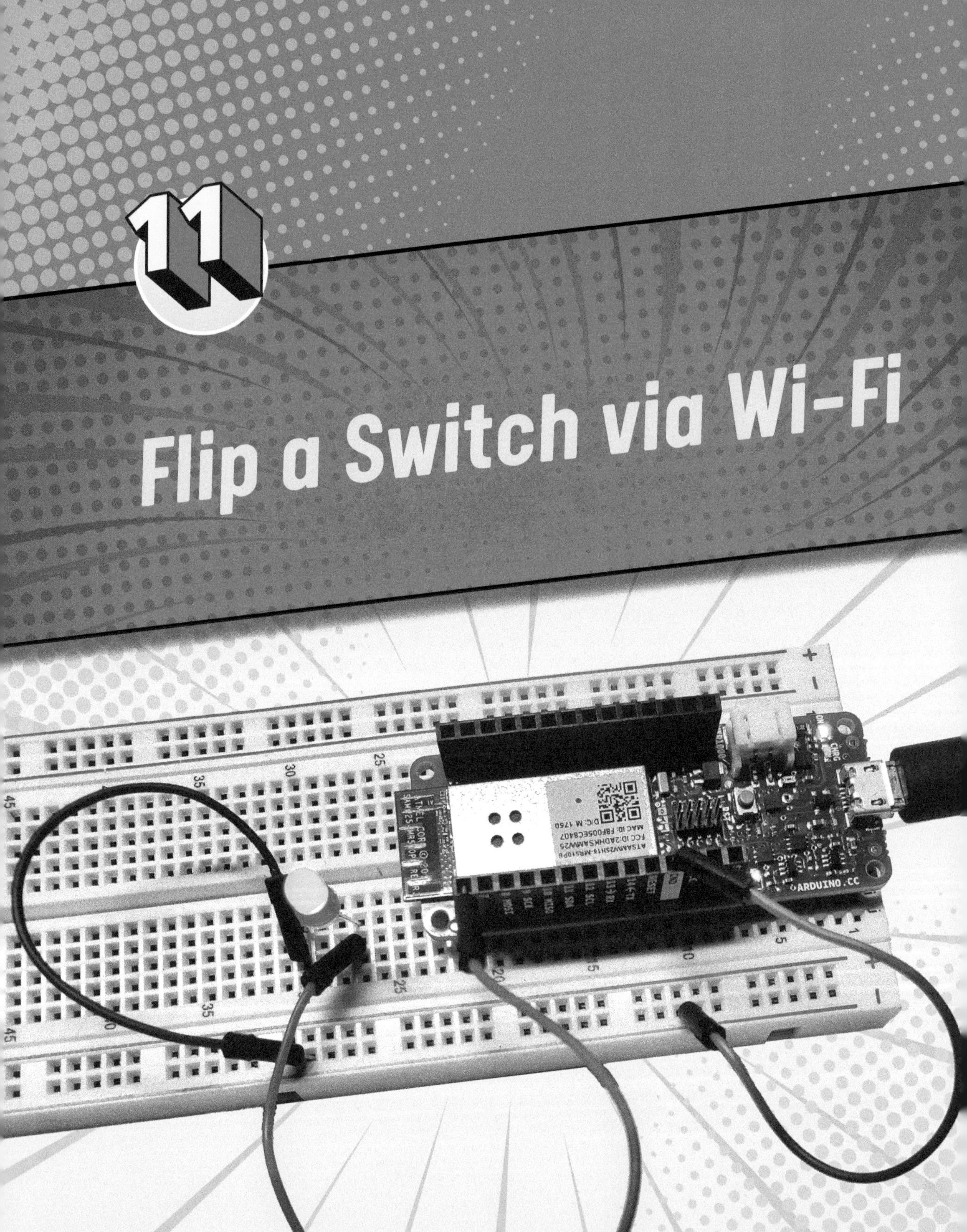
11
Flip a Switch via Wi-Fi
ARDUINO.CC

MATERIALS USED IN CHAPTER 11:

- **Same materials as chapter 10**

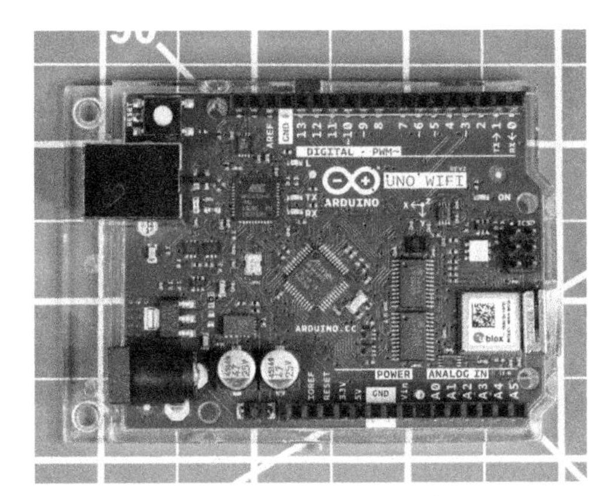

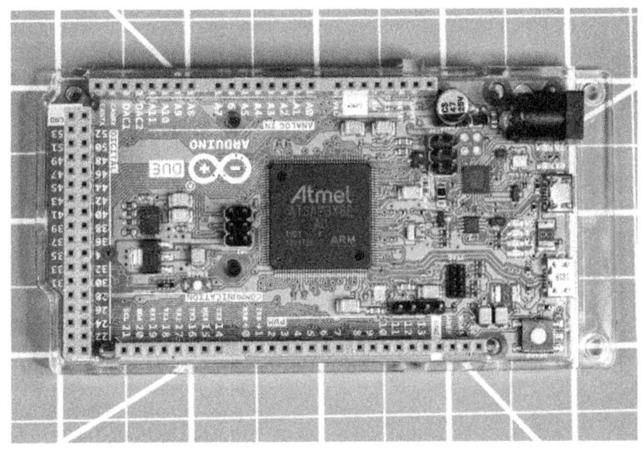

After you have completed the project in this chapter you will have the capacity to control something in your physical environment via Wi-Fi.

The boards that you will be using are the Arduino Uno Wi-Fi and the Arduino Due. Both are Wi-Fi-enabled boards similar to the MKR1000. You can use any of these boards for this project. Since you already have the MKR1000 set up from chapter 10 we will use that, but the same approach will work with any Wi-Fi Arduino board.

This chapter picks up with where we left off in chapter 10. We will simply be blinking an LED via code. The code is significantly more complex, because we are telling the board to set up a web page on our Wi-Fi network and monitor the page for clicks, which will then be used as inputs to the program to switch the light on or off. The code in this chapter is much too complex to type from an example on the page; you need to download this code. The version we are using was modified by Dr. Charif Mahmoudi and can be found by searching for "Arduino MKR1000 Getting Started." There are a lot of versions of this code out there, but this one has the best support for Windows. I have presented a lot of the projects in this book in OS X, but I used the Arduino online via a Chromebook when I built this project.

WIRE THE CIRCUIT

For this project we will be using pin 6 and GND on the MKR1000. The goal here is to simply blink an LED wired between this pin and GND.

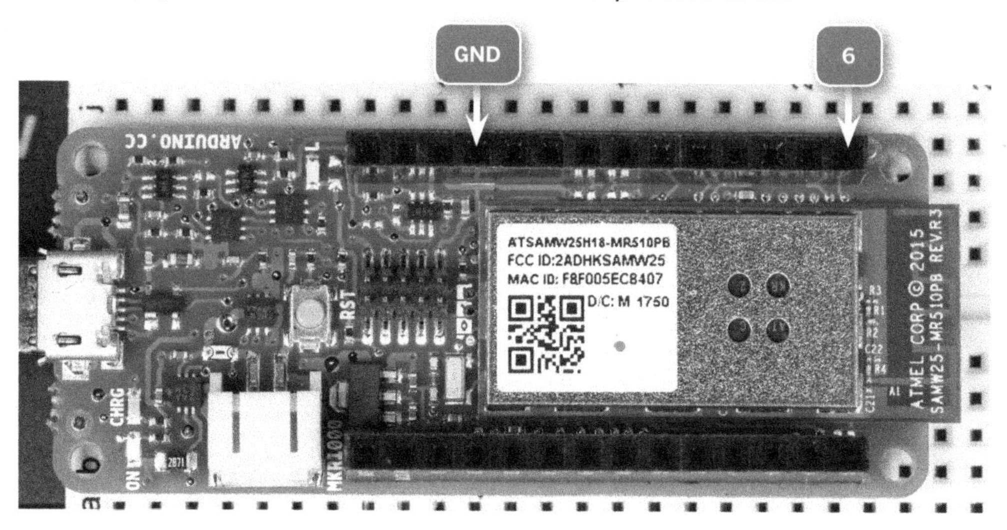

In the image below, you can see pin 6 wired to the long leg of the LED and the GND wired to the short leg of the LED.

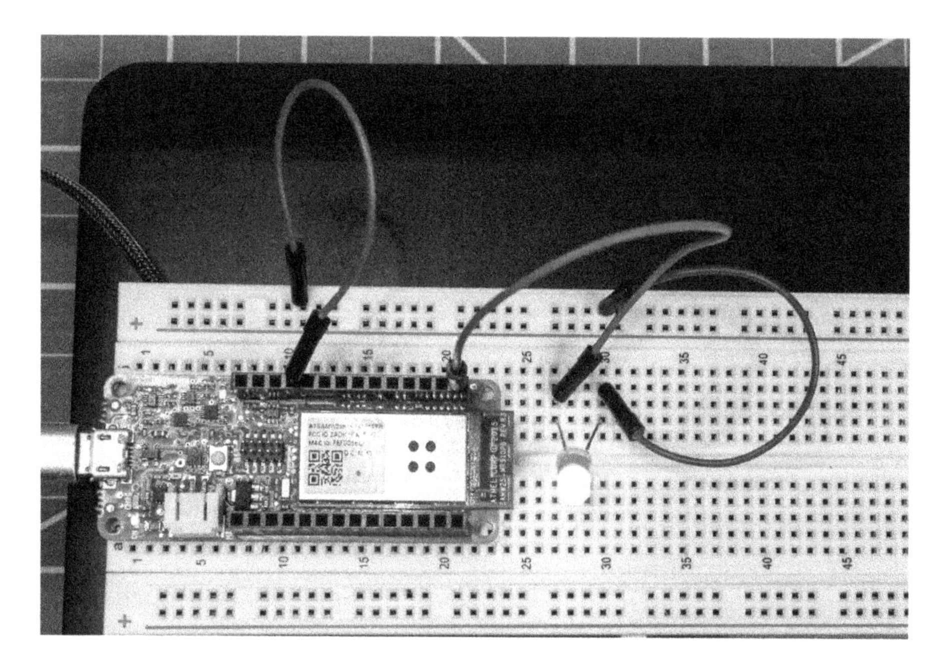

ADD THE WIFI101 LIBRARY

Add the WiFi101 Library with Tools > Manage Libraries. Search for WiFi101 and add the current version (0.16.0 as of this writing).

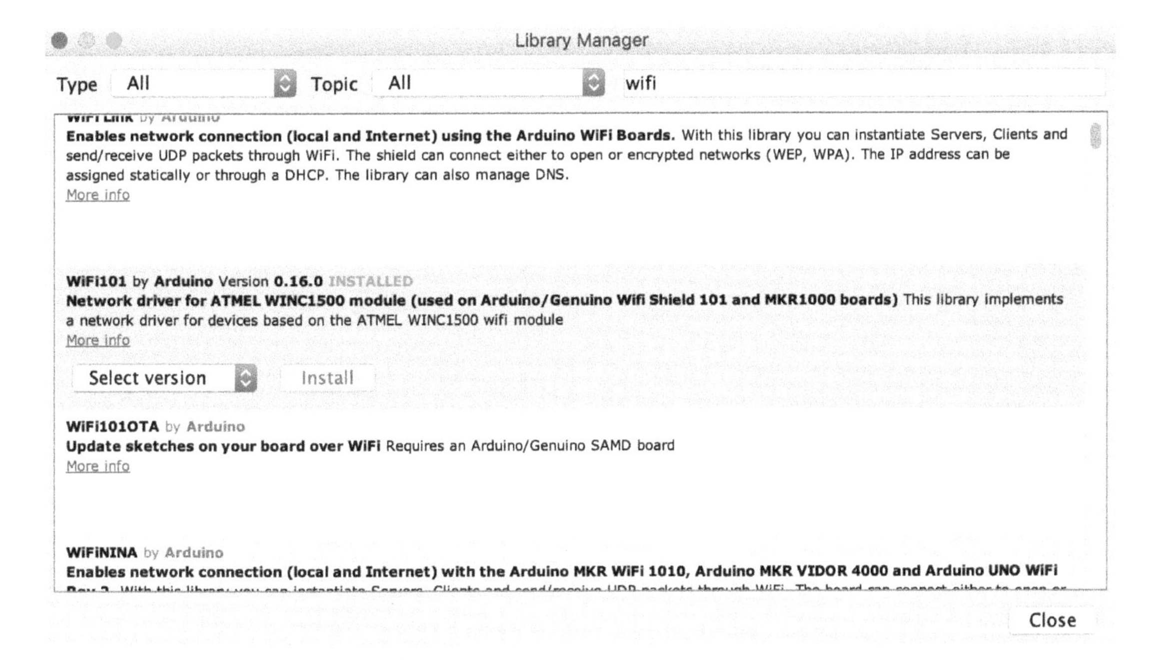

DOWNLOAD THE CODE

The code for this project can be found at MakerAwaker.com under the Action Arduino link or from searching for “Arduino MKR1000 Getting Started.” There are many versions of this code out there.

You need to put your Wifi network name and password here in quotes.

```
#include <WiFi101.h>
#include <WiFiClient.h>
#include <WiFiServer.h>
#include <WiFiSSLClient.h>
#include <WiFiUdp.h>
#include <SPI.h>
#include <WiFi101.h>

char ssid[] =              // your network SSID (name)
char pass[] =              // your network password
int keyIndex = 0;              // your network key Index number (needed only for WEP)
int ledpin = 6;
bool val = true;

int status = WL_IDLE_STATUS;
WiFiServer server(80);

void setup() {
  Serial.begin(9600);      // initialize serial communication
  Serial.print("Start Serial ");
  pinMode(ledpin, OUTPUT);      // set the LED pin mode
  // Check for the presence of the shield
  Serial.print("WiFi101 shield: ");
  if (WiFi.status() == WL_NO_SHIELD) {
    Serial.println("NOT PRESENT");
    return; // don't continue
  }
  Serial.println("DETECTED");
  // attempt to connect to Wifi network:
  while ( status != WL_CONNECTED) {
    digitalWrite(ledpin, LOW);
    Serial.print("Attempting to connect to Network named: ");
    Serial.println(ssid);                   // print the network name (SSID);
    digitalWrite(ledpin, HIGH);
    // Connect to WPA/WPA2 network. Change this line if using open or WEP network:
    status = WiFi.begin(ssid, pass);
    // wait 10 seconds for connection:
    delay(10000);
  }
  server.begin();                           // start the web server on port 80
  printWifiStatus();                        // you're connected now, so print out the status
  digitalWrite(ledpin, HIGH);
}
void loop() {
  WiFiClient client = server.available();   // listen for incoming clients

  if (client) {                             // if you get a client,
    Serial.println("new client");           // print a message out the serial port
    String currentLine = "";                // make a String to hold incoming data from the client
    while (client.connected()) {            // loop while the client's connected
      if (client.available()) {             // if there's bytes to read from the client,
        char c = client.read();             // read a byte, then
        Serial.write(c);                    // print it out the serial monitor
        if (c == '\n') {                    // if the byte is a newline character
```

```
        // if the current line is blank, you got two newline characters in a row.
        // that's the end of the client HTTP request, so send a response:
        if (currentLine.length() == 0) {
          // HTTP headers always start with a response code (e.g. HTTP/1.1 200 OK)
          // and a content-type so the client knows what's coming, then a blank line:
          client.println("HTTP/1.1 200 OK");
          client.println("Content-type:text/html");
          client.println();

          // the content of the HTTP response follows the header:
          client.print("Click <a href=\"/H\">here</a> turn the LED on pin 9 on<br>");
          client.print("Click <a href=\"/L\">here</a> turn the LED on pin 9 off<br>");

          // The HTTP response ends with another blank line:
          client.println();
          // break out of the while loop:
          break;
        }
        else {      // if you got a newline, then clear currentLine:
          currentLine = "";
        }
      }
      else if (c != '\r') {    // if you got anything else but a carriage return character,
        currentLine += c;      // add it to the end of the currentLine
      }

      // Check to see if the client request was "GET /H" or "GET /L":
      if (currentLine.endsWith("GET /H")) {
        digitalWrite(ledpin, HIGH);               // GET /H turns the LED on
      }
      if (currentLine.endsWith("GET /L")) {
        digitalWrite(ledpin, LOW);                // GET /L turns the LED off
      }
    }
  }
  // close the connection:
  client.stop();
  Serial.println("client disonnected");
 }
}

void printWifiStatus() {
  // print the SSID of the network you're attached to:
  Serial.print("SSID: ");
  Serial.println(WiFi.SSID());

  // print your WiFi shield's IP address:
  IPAddress ip = WiFi.localIP();
  Serial.print("IP Address: ");
  Serial.println(ip);

  // print the received signal strength:
  long rssi = WiFi.RSSI();
  Serial.print("signal strength (RSSI):");
  Serial.print(rssi);
  Serial.println(" dBm");
  // print where to go in a browser:
  Serial.print("To see this page in action, open a browser to http://");
  Serial.println(ip);
}
```

MAKE A NEW SKETCH IN THE ARDUINO IDE

Verify and upload the code. Once you have done this, you only need to provide power to the board from a battery, and it will run the code on the Wi-Fi network.

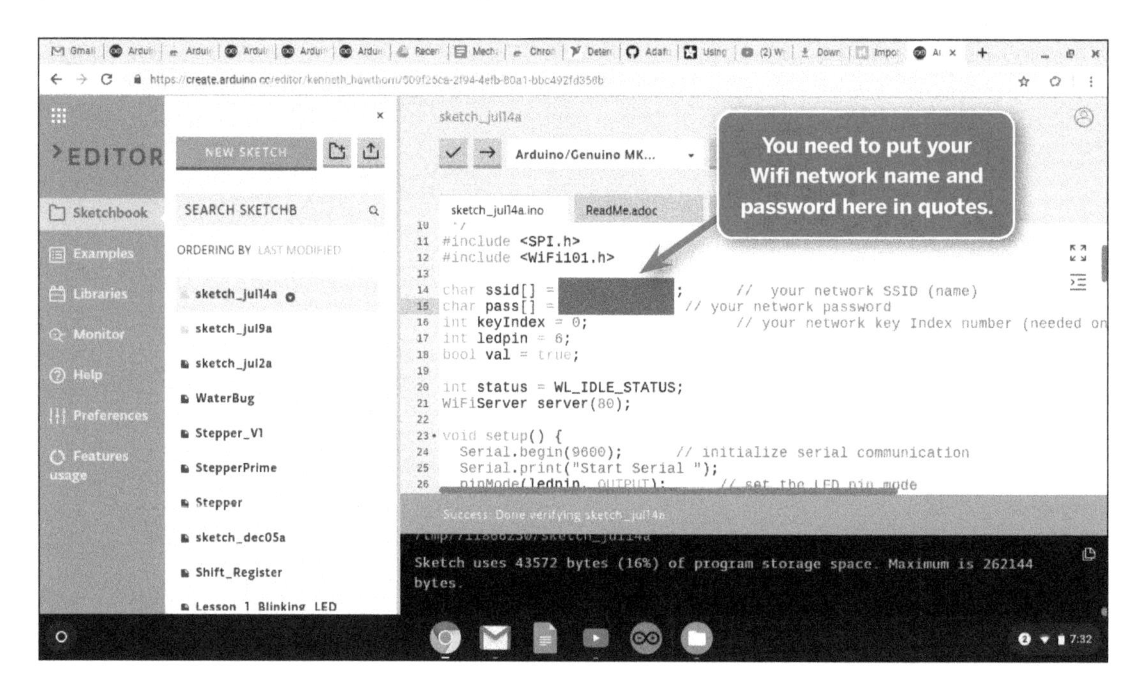

FLIP THE SWITCH!

At this point you only need to go to the address 192.168.232 from any device connected to your Wi-Fi network and click the on or off button, and you will control the LED!

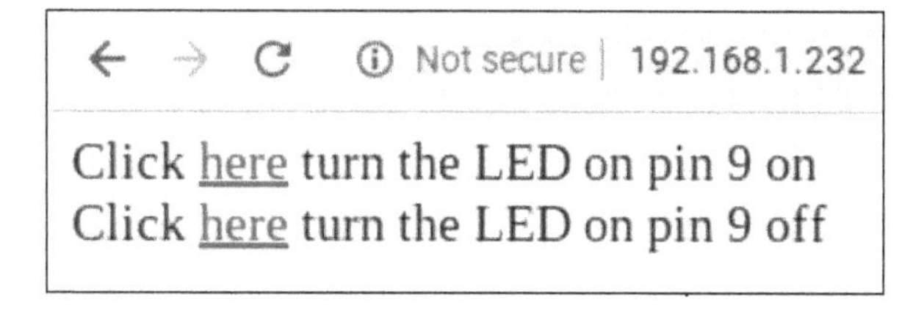

The completed project is shown below.

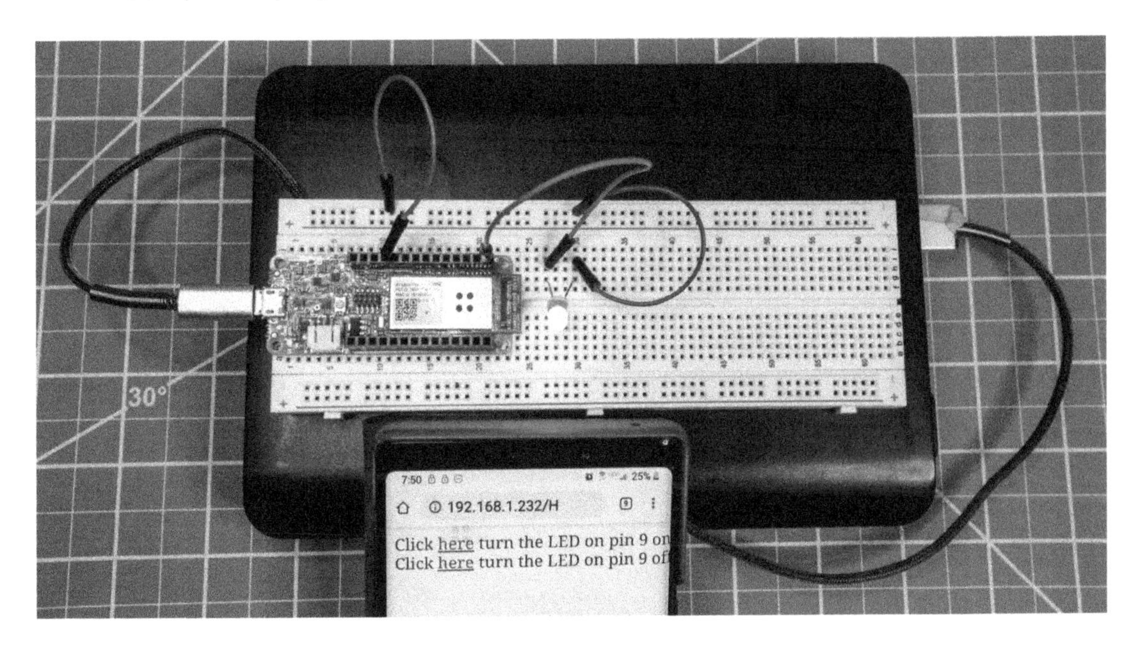

IN THIS CHAPTER YOU:

- completed a full IoT project!
- sent a signal via Wi-Fi to your Arduino to turn on an LED.
- dove into some deepwater code and tweaked it to serve your purpose.

TAKING IT FURTHER

The next step is a baby step. Can you send the signal that turns on the LED in this last project to the IoT relay box? Simply replacing one with the other will allow you to turn lights on and off in your home via your smartphone!

Now that you've completed these projects, you're a certified Arduinist! You have the skills to take any idea you have for a machine and turn it into reality. Make a useful time-saving invention or simply something that will make people smile. Just MAKE it. I suspect the seeds of the idea are already there. You now have the skills to give those ideas physical and functional form. I would really love to see what you build. Send me your work through MakerAwaker.com or ping me on Twitter @ken_hawthorn.

RESOURCES

For more information, please investigate:

Books

Andrews, Warren. *Arduino Playground: Geeky Projects for the Experienced Maker.* No Starch Press, 2017.

Monk, Simon. *Make: Action: Movement, Light, and Sound with Arduino and Raspberry Pi.* Maker Media, Inc., 2016.

Websites

Arduino
arduino.cc
This is the main source for all things Arduino.

Heavy Eyes
heavyeyes.co/shop
A website where you can buy the shark pattern from chapter 5.

Kenneth Hawthorn's blog
MakerAwaker.com

Tinkercad
www.Tinkercad.com
The web page for Arduino circuit simulation.

Other

Twitter account @Ken_Hawthorn
This is the author's twitter account; no politics, just very cool makers of all stripes.

INDEX

ACKNOWLEDGMENTS

I'd like to thank the students of St. Raymond School, Peter Ferrell, Diego Fonstad, Michelle Day, and Peggy Eichman. For Chapter 5, a special thanks to Matt Cavanaugh at Heavy Eyes for opening my eyes to the beautiful world of complex bendable 2-D nets. You have inspired so many students! For Chapter 8, a special thanks to Colin Dixon, Sherry Hsi, and the work of Susan Klimczak and all those who are working to show the power of the ATTINY85 to the non-engineer. I'd like to credit PaperMech.net and Hyunjoo Oh for the paper egg inspiration.

ABOUT THE AUTHOR

Kenneth Hawthorn is a mechanical engineer who became a teacher. As a consulting engineer, he worked on a wide range of original designs from a land-speed record-setting electric motorcycle to LSPR machines. He holds a U.S. Patent for a Fixed Pitch Continuously Variable Transmission. Mr. Hawthorn was selected as the first Salzburg Scholar from the Connie L. Lurie College of Education at San Jose State University where he earned a bachelor's in elementary education. Mr. Hawthorn founded an after-school introductory engineering program that served at-risk students in 17 school districts in the San Jose area for over a decade.

Mr. Hawthorn built the mechatronics program at St. Raymond School in Menlo Park, California, which functioned as a professional development hub for the San Francisco Archdiocese and other local and international schools. He also ran professional development workshops for the Tech Museum in San Jose and the Krause Center for Innovation and has been a presenter at the Minolta Planetarium.

Today, Mr. Hawthorn develops Professional Development and Design Thinking classes at Magellan International School in Austin, Texas. His classes are cross-curricular and blend the five domains of coding, 3-D design, hardware, rapid prototyping, and intellectual property. His students gain familiarity with a wide variety of tools within each of the five domains in order to become technology generalists. Mr. Hawthorn is on the National Teacher Advisory Council for Autodesk and was selected as one of the first ten Dremel Idea Builder Ambassadors.

When he is not traveling and offering professional development workshops to teachers, Mr. Hawthorn lives in Austin, Texas, with his wife, two fluffy cats, and two miniature llamas.

CPSIA information can be obtained
at www.ICGtesting.com
Printed in the USA
BVHW061746131119

9 781641 525992

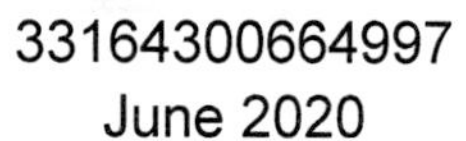
33164300664997
June 2020